CAMBRIDGE LIBRARY COLLECTION

Books of enduring scholarly value

Life Sciences

Until the nineteenth century, the various subjects now known as the life sciences were regarded either as arcane studies which had little impact on ordinary daily life, or as a genteel hobby for the leisured classes. The increasing academic rigour and systematisation brought to the study of botany, zoology and other disciplines, and their adoption in university curricula, are reflected in the books reissued in this series.

Life and Letters of Thomas Henry Huxley

Thomas Henry Huxley (1825–95), the English biologist and naturalist, was known as 'Darwin's Bulldog', and is best remembered today for his vociferous support for Darwin's theory of evolution. He was, however, an influential naturalist, anatomist and religious thinker, who coined the term 'agnostic' to describe his own beliefs. Almost entirely self-educated, he became an authority in anatomy and palaeontology, and after the discovery of the archaeopteryx, he was the first to suggest that birds had evolved from dinosaurs. He was also a keen promoter of scientific education who strove to make science a paid profession, not dependent on patronage or wealth. Published in 1903, this three-volume work, compiled by his son Leonard Huxley, is the second and most complete edition of Huxley's biography and selected letters. Volume 3 covers the period 1887–95, during which Huxley battled ill-health while continuing to defend his scientific ideals.

Cambridge University Press has long been a pioneer in the reissuing of out-of-print titles from its own backlist, producing digital reprints of books that are still sought after by scholars and students but could not be reprinted economically using traditional technology. The Cambridge Library Collection extends this activity to a wider range of books which are still of importance to researchers and professionals, either for the source material they contain, or as landmarks in the history of their academic discipline.

Drawing from the world-renowned collections in the Cambridge University Library, and guided by the advice of experts in each subject area, Cambridge University Press is using state-of-the-art scanning machines in its own Printing House to capture the content of each book selected for inclusion. The files are processed to give a consistently clear, crisp image, and the books finished to the high quality standard for which the Press is recognised around the world. The latest print-on-demand technology ensures that the books will remain available indefinitely, and that orders for single or multiple copies can quickly be supplied.

The Cambridge Library Collection brings back to life books of enduring scholarly value (including out-of-copyright works originally issued by other publishers) across a wide range of disciplines in the humanities and social sciences and in science and technology.

Life and Letters of Thomas Henry Huxley

VOLUME 3

LEONARD HUXLEY

CAMBRIDGE UNIVERSITY PRESS

Cambridge, New York, Melbourne, Madrid, Cape Town,
Singapore, São Paolo, Delhi, Tokyo, Mexico City

Published in the United States of America by Cambridge University Press, New York

www.cambridge.org
Information on this title: www.cambridge.org/9781108040488

© in this compilation Cambridge University Press 2012

This edition first published 1903
This digitally printed version 2012

ISBN 978-1-108-04048-8 Paperback

LIFE AND LETTERS

OF

THOMAS HENRY HUXLEY

M? Queen, Sc.

T. H. Huxley.

From a photograph by Donney, 1890.

Life and Letters

OF

Thomas Henry Huxley

BY HIS SON

LEONARD HUXLEY

IN THREE VOLUMES

VOL. III

London

MACMILLAN AND CO., Limited

NEW YORK: THE MACMILLAN COMPANY

1903

First Edition, 2 vols. 8vo, October 1900.
Reprinted November and December 1900.
Second Edition, Globe 8vo (Eversley Series, 3 vols.), 1903.

CONTENTS

CHAPTER I

1887

THE first half of 1887, like that of the preceding year, was chequered by constant returns of ill-health. "As one gets older," he writes in a New Year's letter to Sir J. Donnelly, "hopes for oneself get more moderate, and I shall be content if next year is no worse than the last. Blessed are the poor in spirit!" The good effects of the visit to Arolla had not outlasted the winter, and from the end of February he was obliged to alternate between London and the Isle of Wight.

Nevertheless, he managed to attend to a good deal of business in the intervals between his periodic flights to the country, for he continued to serve on the Royal Society Council, to do some of the examining work at South Kensington, and to fight for the establishment of adequate Technical Education in England. He attended the Senate and various committees of the London University and of the Marine Biological Association.

Several letters refer to the proposal—it was the Jubilee year—to commemorate the occasion by the establishment of the Imperial Institute. To this he gladly gave his support; not indeed to the merely social side; but in the opportunity of organising the practical applications of science to industry he saw the key to success in the industrial war of the future. Seconding the resolution proposed by Lord Rothschild at the Mansion House meeting on January 12, he spoke of the relation of industry to science—the two great developments of this century. Formerly practical men looked askance at science, " but within the last thirty years, more particularly," continues the report in *Nature* (vol. xxxiii. p. 265) "that state of things had entirely changed. There began in the first place a slight flirtation between science and industry, and that flirtation had grown into an intimacy, he might almost say courtship, until those who watched the signs of the times saw that it was high time that the young people married and set up an establishment for themselves. This great scheme, from his point of view, was the public and ceremonial marriage of science and industry."

Proceeding to speak of the contrast between militarism and industrialism, he asked whether, after all, modern industry was not war under the forms of peace. The difference was the difference between modern and ancient war, consisting in the use of scientific weapons, of organisation and information. The country, he concluded, had dropped astern in the race for want of special education which was obtained elsewhere by the artisan. The only possible chance for keeping the industry of England at the head of the world was through organisation.

Writing on January 18, to Mr. Herbert Spencer, who had sent him some proofs of his Autobiography to look through, he says :—

I see that your proofs have been in my hands longer than I thought for. But you may have seen that I have been "starring" at the Mansion House.

This was not exactly one of those bits of over-easiness to pressure with which you reproach me—but the resultant of a composition of pressures, one of which was the conviction that the "Institute" might be made into something very useful and greatly wanted—if only the projectors could be made to believe that they had always intended to do that which your humble servant wants done—that is the establishment of a sort of Royal Society for the improvement of industrial knowledge and an industrial university—by voluntary association.

I hope my virtue may be its own reward. For except being knocked up for a day or two by the unwonted effort, I doubt whether there will be any other. The thing has fallen flat as a pancake, and I greatly doubt whether any good will come of it. Except a fine in the shape of a subscription, I hope to escape further punishment for my efforts to be of use.

However, this was only the beginning of his campaign.

On January 27, a letter from him appeared in the *Times*, guarding against a wrong interpretation of his speech, in the general uncertainty as to the intentions of the proposers of the scheme.

I had no intention (he writes) of expressing any enthusiasm on behalf of the establishment of a vast permanent bazaar.   I am not competent to estimate the real utility of these great shows.   What I do see very clearly is that they involve difficulties of site, huge working expenses, the potentiality of endless squabbles, and apparently the cheapening of knighthood.

As for the site proposed at South Kensington, "the arguments used in its favour in the report would be conclusive if the dry light of reason were the sole guide of human action."   But it would alienate other powerful and wealthy bodies, which were interested in the Central Institute of the City and Guilds Technical Institute, "which looks so portly outside and is so very much starved inside."

He wrote again to the *Times* on March 21 :—

The Central Institute is undoubtedly a splendid monument of the munificence of the city.   But munificence without method may arrive at results indistinguishably similar to those of stinginess.   I have been blamed for saying that the Central Institute is "starved."   Yet a man who has only half as much food as he needs is indubitably starved, even though his short rations consist of ortolans and are served upon gold plate.

Only half the plan of operations as drawn up by the Committee was, or could be, carried out on existing funds.

The later part of his letter was printed by the Committee as defining the functions of the new Institute :—

That with which I did intend to express my strong sympathy was the intention which I thought I discerned

to establish something which should play the same part
in regard to the advancement of industrial knowledge
which has been played in regard to science and learning
in general, in these realms, by the Royal Society and the
Universities. . . . I pictured the Imperial Institute to
myself as a house of call for all those who are concerned
in the advancement of industry ; as a place in which the
home-keeping industrial could find out all he wants to
know about colonial industry and the colonist about home
industry ; as a sort of neutral ground on which the
capitalist and the artisan would be equally welcome ; as
a centre of intercommunication in which they might enter
into friendly discussion of the problems at issue between
them, and, perchance, arrive at a friendly solution of them.
I imagined it a place in which the fullest stores of industrial knowledge would be made accessible to the
public ; in which the higher questions of commerce and
industry would be systematically studied and elucidated ;
and where, as in an industrial university, the whole
technical education of the country might find its centre and
crown.   If I earnestly desire to see such an institution
created, it is not because I think that or anything else
will put an end to pauperism and want—as somebody
has absurdly suggested,—but because I believe it will
supply a foundation for that scientific organisation of our
industries which the changed conditions of the times
render indispensable to their prosperity.   I do not think
I am far wrong in assuming that we are entering, indeed,
have already entered, upon the most serious struggle for
existence to which this country has ever been committed.
The latter years of the century promise to see us em-
barked in an industrial war of far more serious import
than the military wars of its opening years.   On the east,
the most systematically instructed and best-informed people
in Europe are our competitors ; on the west, an energetic
offshoot of our own stock, grown bigger than its parent,
enters upon the struggle possessed of natural resources to

which we can make no pretension, and with every prospect
of soon possessing that cheap labour by which they may be
effectually utilised.   Many circumstances tend to justify
the hope that we may hold our own if we are careful to
" organise victory."   But to those who reflect seriously on
the prospects of the population of Lancashire and Yorkshire—should the time ever arrive when the goods which
are produced by their labour and their skill are to be had
cheaper elsewhere—to those who remember the cotton
famine and reflect how much worse a customer famine
would be, the situation appears very grave.

On February 19 and 22, he wrote again to the
Times declaring against the South Kensington site.
It was too far from the heart of commercial
organisation in the city, and the city people were
preparing to found a similar institution of their own.
He therefore wished to prevent the Imperial Institute
from becoming a weak and unworthy memorial of the
reign.

A final letter to the *Times* on March 21, was
evoked by the fact that Lord Hartington, in giving
away the prizes at the Polytechnic Y.M.C.A., had
adopted Huxley's position as defined in his speech,
and declared that science ought to be aided on precisely the same grounds on which we aid the army
and navy.

In this letter he asks, how do we stand prepared
for the task thus imperatively set us ?   We have the
machinery for providing instruction and information,
and for catching capable men, but both in a
disjointed condition—" all mere torsos—fine, but

fragmentary." "The ladder from the School Board to the Universities, about which I dreamed dreams many years ago, has not yet acquired much more substantiality than the ladder of Jacob's vision," but the Science and Art Department, the Normal School of Science, and the Central Institute only want the means to carry out the recommendations already made by impartial and independent authority. "Economy does not lie in sparing money, but in spending it wisely."

He concluded with an appeal to Lord Hartington to take up this task of organising industrial education and bring it to a happy issue.

A proposal was also made to the Royal Society to co-operate, and Sir M. Foster writes on February 19 : " We have appointed a Committee to consider and draw up a draft reply with a view of the R.S. following up your letter."

To this Huxley replied on the 22nd :—

. . . My opinion is that the R.S. has no right to spend its money or pledge its credit for any but scientific objects, and that we have nothing to do with sending round the hat for other purposes.

The project of the Institute Committee as it stands connected with the South Kensington site—is condemned by all the city people and will receive none but the most grudging support from them. They are going to set up what will be practically an Institute of their own in the city.

The thing is already a failure. I daresay it will go on and be varnished into a simulacrum of success—to become eventually a ghost like the Albert Hall or revive as a tea garden.

The following letter also touches upon the function
of the Institute from the commercial side :—

4 MARLBOROUGH PLACE,
Feb. 20, 1887.

MY DEAR DONNELLY—Mr. Law's suggestion gives
admirable definition to the notions that were floating in
my mind when I wrote in my letter to the *Times*, that I
imagined the Institute would be "a place in which the
fullest stores of industrial knowledge would be made
accessible to the public." A man of business who wants to
know anything about the prospects of trade with say,
Borrioboola-Gha (*vide* Bleak House) ought to be able to
look into the Institute and find there somebody who will
at once fish out for him among the documents in the
place all that is known about Borrioboola.

But a Commercial Intelligence Department is not all
that is wanted, *vide* valuable letter aforesaid.

I hope your appetite for the breakfast was none the
worse for last night's doings—mine was rather improved,
but I am dog-tired.—Ever yours very faithfully,

T. H. HUXLEY.

I return Miss ——'s note. She evidently thinks my
cage is labelled "These animals bite."

Later in the year, the following letters show him
continuing the campaign. But an attack of pleurisy,
which began the very day of the Jubilee, prevented
him from coming to speak at a meeting upon
Technical Education. In the autumn, however, he
spoke on the subject at Manchester, and had the
satisfaction of seeing the city "go solid," as he
expressed it, for technical education. The circumstances of this visit are given later.

4 MARLBOROUGH PLACE, N.W.,
May 1, 1887.

MY DEAR ROSCOE—I met Lord Hartington at the
Academy Dinner last night and took the opportunity of
urging upon him the importance of following up his
technical education speech.   He told me he had been in
communication with you about the matter, and he seemed
to me to be very well disposed to your plans.

I may go on crying in the wilderness until I am
hoarse, with no result, but if he and you and Mundella
will take it up, something may be done.—Ever yours
very faithfully,   T. H. HUXLEY.

4 MARLBOROUGH PLACE,
June 28, 1887.

MY DEAR ROSCOE—Donnelly was here on Sunday and
was quite right up to date.   I felt I ought to be better,
and could not make out why the deuce I was not.
Yesterday the mischief came out.   There is a touch of
pleurisy—which has been covered by the muscular
rheumatism.

So I am relegated to bed and told to stop there—with
the company of cataplasms to keep me lively.

I do not think the attack in any way serious—but
M. Pl. is a gentleman not be trifled with, when you are over
sixty, and there is nothing for it but to obey my doctor's
orders.

Pray do not suppose I would be stopped by a trifle, if
my coming to the meeting [1] would really have been of use.
I hope you will say how grieved I am to be absent.—
Ever yours very faithfully,   T. H. HUXLEY.

---

[1] Of July 1, on Technical Education.

4 MARLBOROUGH PLACE,
June 29, 1887.

MY DEAR ROSCOE—I have scrawled a variety of comments on the paper you sent me. Deal with them as you think fit.

Ever since I was on the London School Board I have seen that the key of the position is in the Sectarian Training Colleges and that wretched imposture, the pupil teacher system. As to the former *Delendae sunt* no truce or pact to be made with them, either Church or Dissenting. Half the time of their students is occupied with grinding into their minds their tweedle-dum and tweedle-dee theological idiocies, and the other half in cramming them with boluses of other things to be duly spat out on examination day. Whatever is done do not let us be deluded by any promises of theirs to hook on science or technical teaching to their present work.

I am greatly disgusted that I cannot come to Tyndall's dinner to-night [1]—but my brother-in-law's death would have stopped me (the funeral to-day)—even if my doctor had not forbidden me to leave my bed. He says I have some pleuritic effusion on one side and must mind my P's and Q's.—Ever yours very faithfully,   T. H. HUXLEY.

A good deal of correspondence at this time with Sir M. Foster relates to the examinations of the Science and Art Department. He was still Dean, it will be remembered, of the Royal College of Science, and further kept up his connection with the Department by acting in an honorary capacity as Examiner, setting questions, but less and less looking over papers, acting as the channel for official communications, as when he writes (April 24), "I send you

[1] See p. 25.

some Department documents—nothing alarming, only more worry for the Asst. Examiners, and that *we* do not mind"; and finally signing the Report. But to do this after taking so small a share in the actual work of examining, grew more and more repugnant to him, till on October 12 he writes :—

> I will read the Report and sign it if need be—though there really must be some fresh arrangement.
> Of course I have entire confidence in your judgment about the examination, but I have a mortal horror of putting my name to things I do not know of my own knowledge.

In addition to these occupations, he wrote a short paper upon a fossil, Ceratochelys, which was read at the Royal Society on March 31; while on April 7 he read at the Linnean (*Botany:* vol. xxiv. pp. 101-124), his paper, "The Gentians : Notes and Queries," which had sprung from his holiday amusement at Arolla.

Philosophy, however, claimed most of his energies. The campaign begun in answer to the incursion of Mr. Lilly was continued in the article "Science and Pseudo-Scientific Realism" (*Coll. Essays,* v. 59-89) which appeared in the *Nineteenth Century* for February 1887. The text for this discourse was the report of a sermon by Canon Liddon, in which that eminent preacher spoke of catastrophes as the antithesis of physical law, yet possible inasmuch as a "lower law" may be "suspended" by the "intervention of a higher," a mode of reasoning which he applied to the possibility of miracles such as that of Cana.

The man of science was up in arms against this incarnation of abstract terms, and offered a solemn protest against that modern recrudescence of ancient realism which speaks of "laws of nature" as though they were independent entities, agents, and efficient causes of that which happens, instead of simply our name for observed successions of facts.

Carefully as all personalities had been avoided in this article, it called forth a lively reply from the Duke of Argyll, rebuking him for venturing to criticise the preacher, whose name was now brought forward for the first time, and raising a number of other questions, philosophical, geological, and biological, to which Huxley rejoined with some selections from the authentic history of these points in "Science and Pseudo-Science" (*Nineteenth Century*, April 1887, *Coll. Essays*, v. 90-125).

Moreover, judging from the vivacity of the Duke's reply that some of the shafts of the first article must have struck nearer home than the pulpit of St. Paul's, he was induced to read " The Reign of Law," the second chapter of which, dealing with the nature of "Law," he now criticised sharply as "a sort of 'summa' of pseudo-scientific philosophy," with its confusions of law and necessity, law and force, "law in the sense, not merely of a rule, but of a cause." [1]

[1] Cf. his treatment of the subject 24 years before, vol. i. p. 349.

He wound up with some banter upon the Duke's picture of a scientific Reign of Terror, whereby, it seemed, all men of science were compelled to accept the Darwinian faith, and against which Huxley himself was preparing to rebel, as if,

forsooth, I am supposed to be waiting for the signal of "revolt," which some fiery spirits among these young men are to raise before I dare express my real opinions concerning questions about which we older men had to fight in the teeth of fierce public opposition and obloquy — of something which might almost justify even the grandiloquent epithet of a Reign of Terror—before our excellent successors had left school.

Here for a while the debate ceased. But in the September number of the *Nineteenth Century* the Duke of Argyll returned to the fray with an article called "A Great Lesson," in which he attempted to offer evidence in support of his assertions concerning the scientific reign of terror. The two chief pieces of evidence adduced were Bathybius and Dr. (now Sir J.) Murray's theory of coral reefs. The former was instanced as a blunder due to the desire of finding support for the Darwinian theory in the existence of this widespread primordial life ; the latter as a case in which a new theory had been systematically burked, for fear of damaging the infallibility of Darwin, who had propounded a different theory of coral reefs !

Huxley's reply to this was contained in the latter half of an article which appeared in the *Nineteenth*

Century for November 1887, under the title of
"Science and the Bishops" (reprinted both in *Contro-
verted Questions* and in the *Collected Essays*, v. 126, as
"An Episcopal Trilogy"). Preaching at Manchester
this autumn, during the meeting of the British
Association, the Bishops of Carlisle, Bedford, and
Manchester had spoken of science not only with
knowledge, but in the spirit of equity and generosity.
"These sermons," he exclaims, "are what the Germans
call Epochemachend!"

> How often was it my fate (he continues), a quarter of
> a century ago, to see the whole artillery of the pulpit
> brought to bear upon the doctrine of evolution and its
> supporters! Any one unaccustomed to the amenities of
> ecclesiastical controversy would have thought we were too
> wicked to be permitted to live.

After thus welcoming these episcopal advances, he
once more repudiated the *à priori* argument against
the efficacy of prayer, the theme of one of the three
sermons, and then proceeded to discuss another
sermon of a dignitary of the Church, which had been
sent to him by an unknown correspondent, for "there
seems to be an impression abroad—I do not desire to
give any countenance to it—that I am fond of reading
sermons."

Now this preacher was of a very different mind
from the three bishops. Instead of dwelling upon
the "supreme importance of the purely spiritual in
our faith," he warned his hearers against dropping
off any of the miraculous integument of their religion.

"Christianity is essentially miraculous, and falls to the ground if miracles be impossible." He was uncompromisingly opposed to any accommodation with advancing knowledge, or with the high standard of veracity, enforced by the nature of their pursuits, in which Huxley found the only difference between scientific men and any other class of the community.

But it was not merely this misrepresentation of science on its speculative side which Huxley deplored; he was roused to indignation by an attack on its morality. The preacher reiterated the charge brought forward in the "Great Lesson," that Dr. Murray's theory of coral reefs had been actually suppressed for two years, and that by the advice of those who accepted it, for fear of upsetting the infallibility of the great master.

Hereupon he turned in downright earnest upon the originator of the assertion, who, he considered, had no more than the amateur's knowledge of the subject. A plain statement of the facts was refutation enough. The new theories, he pointed out, had been widely discussed; they had been adopted by some geologists, although Darwin himself had not been converted, and after careful and prolonged re-examination of the question, Professor Dana, the greatest living authority on coral reefs, had rejected them. As Professor Judd said, "If this be a 'conspiracy of silence,' where, alas! can the geological speculator seek for fame?" Any warning not to publish in haste was but advice to a still unknown man not to attack a seemingly well-established theory without making sure of his ground.[1]

As for the Bathybius myth, Huxley pointed out that his announcement of the discovery had been simply a statement of the actual facts, and that so far from seeing in it a confirmation of Darwinian hypotheses, he was careful to warn his readers "to keep the questions of fact and the questions of interpretation well apart." " That which interested me in the matter," he says, " was the apparent analogy of *Bathybius* with other well-known forms of lower life," . . . " if *Bathybius* were brought up alive from the bottom of the Atlantic to-morrow, the fact would not have the slightest bearing, that I can discern, upon Mr. Darwin's speculations, or upon any of the disputed problems of biology." And as for his " eating the leek " afterwards, his ironical account of it is an instance of how the adoption of a plain, straightforward course can be described without egotism.

> The most considerable difference I note among men (he concludes) is not in their readiness to fall into error, but in their readiness to acknowledge these inevitable lapses.

As the Duke in a subsequent article did not unequivocally withdraw his statements, Huxley declined to continue public controversy with him.

Three years later, writing (October 10, 1890) to

---

[1] Letter in *Nature.*

Sir J. Donnelly apropos of an article by Mr. Mallock in the *Nineteenth Century*, which made use of the "Bathybius myth," he says :—

Bathybius is far too convenient a stick to beat this dog with to be ever given up, however many lies may be needful to make the weapon effectual.

I told the whole story in my reply to the Duke of Argyll, but of course the pack give tongue just as loudly as ever. Clerically-minded people cannot be accurate, even the liberals.

I give here the letter sent to the "unknown correspondent" in question, who had called his attention to the fourth of these sermons.

4 MARLBOROUGH PLACE,
Sept. 30, 1887.

I have but just returned to England after two months' absence, and in the course of clearing off a vast accumulation of letters, I have come upon yours.

The Duke of Argyll has been making capital out of the same circumstances as those referred to by the Bishop. I believe that the interpretation put upon the facts by both is wholly misleading and erroneous.

It is quite preposterous to suppose that the men of science of this or any other country have the slightest disposition to support any view which may have been enunciated by one of their colleagues, however distinguished, if good grounds are shown for believing it to be erroneous.

When Mr. Murray arrived at his conclusions I have no doubt he was advised to make his ground sure before he attacked a generalisation which appeared so well founded as that of Mr. Darwin respecting coral reefs.

If he had consulted me I should have given him that

advice myself, for his own sake.   And whoever advised
him, in that sense, in my opinion did wisely.

But the theologians cannot get it out of their heads,
that as they have creeds, to which they must stick at all
hazards, so have the men of science.   There is no more
ridiculous delusion.   We, at any rate, hold ourselves
morally bound to "try all things and hold fast to that
which is good "; and among public benefactors, we reckon
him who explodes old error, as next in rank to him who
discovers new truth.

You are at liberty to make any use you please of this
letter.

Two letters on kindred subjects may appropriately
follow in this place.   Thanking M. Henri Gadeau de
Kerville for his "Causeries sur le Transformisme," he
writes (Feb. 1) :—

DEAR SIR—Accept my best thanks for your interesting
"causeries," which seem to me to give a very clear view
of the present state of the evolution doctrine as applied
to biology.

There is a statement on p. 87 "Après sa mort Lamarck
fut complètement oublié," which may be true for France
but certainly is not so for England.   From 1830 onwards
for more than forty years Lyell's "Principles of Geology"
was one of the most widely read scientific books in this
country, and it contains an elaborate criticism of Lamarck's
views.   Moreover, they were largely debated during the
controversies which arose out of the publication of the
"Vestiges of Creation" in 1844 or thereabouts.   We are
certainly not guilty of any neglect of Lamarck on this
side of the Channel.

If I may make another criticism it is that, to my
mind, atheism is, on purely philosophical grounds, untenable.   That there is no evidence of the existence of
such a being as the God of the theologians is true enough ;

but strictly scientific reasoning can take us no further. Where we know nothing we can neither affirm nor deny with propriety.

The other is in answer to the Bishop of Ripon, enclosing a few lines on the principal representatives of modern science, which he had asked for.

> 4 MARLBOROUGH PLACE
> *June* 16, 1887.

MY DEAR BISHOP OF RIPON—I shall be very glad if I can be of any use to you now and always. But it is not an easy task to put into half-a-dozen sentences, up to the level of your vigorous English, a statement that shall be unassailable from the point of view of a scientific fault-finder—which shall be intelligible to the general public and yet accurate.

I have made several attempts and enclose the final result. I think the substance is all right, and though the form might certainly be improved, I leave that to you. When I get to a certain point of tinkering my phrases I have to put them aside for a day or two.

Will you allow me to suggest that it might be better not to name any living man? The temple of modern science has been the work of many labourers not only in our own but in other countries. Some have been more busy in shaping and laying the stones, some in keeping off the Sanballats, some prophetwise in indicating the course of the science of the future. It would be hard to say who has done best service. As regards Dr. Joule, for example, no doubt he did more than any one to give the doctrine of the conservation of energy precise expression, but Mayer and others run him hard.

Of deceased Englishmen who belong to the first half of the Victorian epoch, I should say that Faraday, Lyell, and Darwin had exerted the greatest influence, and all three were models of the highest and best class of physical philosophers.

As for me, in part from force of circumstance and in part from a conviction I could be of most use in that way, I have played the part of something between maid-of-all-work and gladiator-general for Science, and deserve no such prominence as your kindness has assigned to me. —With our united kind regards to Mrs. Carpenter and yourself, ever yours very faithfully,   T. H. HUXLEY.

A brief note, also, to Lady Welby, dated July 25, is characteristic of his attitude towards unverified speculation.

I have looked through the paper you have sent me, but I cannot undertake to give any judgment upon it. Speculations such as you deal with are quite out of my way. I get lost the moment I lose touch of valid fact and incontrovertible demonstration and find myself wandering among large propositions, which may be quite true but which would involve me in months of work if I were to set myself seriously to find out whether, and in what sense, they are true. Moreover, at present, what little energy I possess is mortgaged to quite other occupations.

The following letter was in answer to a request which I was commissioned to forward him, that he would consent to serve on an honorary committee of the Société des Professeurs de Français en Angleterre.

Jan. 17, 1887.

I quite forgot to say anything about the Comité d'honneur, and as you justly remark in the present strained state of foreign politics the consequences may be serious. Please tell your colleague that I shall be " proud an' 'appy." You need not tell him that my pride and happiness are contingent on having nothing to do for the honour.

In the meantime, the ups and downs of his health are reflected in various letters of these six months. Much set up by his stay in the Isle of Wight, he writes from Shanklin on April 11 to Sir E. Frankland, describing the last meeting of the *x* Club, which the latter had not been able to attend, as he was staying in the Riviera :—

> Hooker, Tyndall, and I alone turned up last Thursday. Lubbock had gone to High Elms about used up by the House of Commons, and there was no sign of Hirst.
>
> Tyndall seemed quite himself again. In fact, we three old fogies voted unanimously that we were ready to pit ourselves against any three youngsters of the present generation in walking, climbing, or head-work, and give them odds.
>
> I hope you are in the same comfortable frame of mind.
>
> I had no notion that Mentone had suffered so seriously in the earthquake of 1887. Moral for architects : read your Bible and build your house upon the rock.
>
> The sky and sea here may be fairly matched against Mentone or any other of your Mediterranean places. Also the east wind, which has been blowing steadily for ten days, and is nearly as keen as the Tramontana. Only in consequence of the long cold and drought not a leaf is out.

Shanklin, indeed, suited him so well that he had half a mind to settle there. "There are plenty of sites for building," he writes home in February, "but I have not thought of commencing a house yet." However, he gave up the idea; Shanklin was too far from town.

But though he was well enough as long as he kept out of London, a return to his life there was not possible for any considerable time. On May 19, just before a visit to Mr. F. Darwin at Cambridge, I find that he went down to St. Albans for a couple of days, to walk; and on the 27th he betook himself, terribly ill and broken down, to the Savernake Forest Hotel, in hopes of getting "screwed up." This "turned out a capital speculation, a charming spick-and-span little country hostelry with great trees in front." But the weather was persistently bad, "the screws got looser rather than tighter," and again he was compelled to stay away from the x.

A week later, however, he writes :—

The weather has been detestable, and I got no good till yesterday, which was happily fine. Ditto to-day, so I am picking up, and shall return to-morrow, as, like an idiot as I am, I promised to take the chair at a public meeting about a Free Library for Marylebone on Tuesday evening.

I wonder if you know this country. I find it charming.

On the same day as that which was fixed for the meeting in favour of the Free Library, he had a very interesting interview with the Premier, of which he left the following notes, written at the Athenæum immediately after :—

June 7, 1887.

Called on Lord Salisbury by appointment at 3 P.M., and had twenty minutes' talk with him about the "matter of some public interest" mentioned in his letter of the (29th).

This turned out to be a proposal for the formal re-cognition of distinguished services in Science, Letters, and Art by the institution of some sort of order analogous to the *Pour le Mérite.* Lord Salisbury spoke of the anomalous present mode of distributing honours, intimated that the Queen desired to establish a better system, and asked my opinion.

I said that I should like to separate my personal opinion from that which I believed to obtain among the majority of scientific men ; that I thought many of the latter were much discontented with the present state of affairs, and would highly approve of such a proposal as Lord Salisbury shadowed forth.

That, so far as my own personal feeling was concerned, it was opposed to anything of the kind for Science.   I said that in Science we had two advantages—first, that a man's work is demonstrably either good or bad ; and secondly, that the "contemporary posterity" of foreigners judges us, and rewards good work by membership of Academies and so forth.

In Art, if a man chooses to call Raphael a dauber, you can't prove he is wrong ; and literary work is just as hard to judge.

I then spoke of the dangers to which science is exposed by the undue prominence and weight of men who suc-cessfully apply scientific knowledge to practical purposes —engineers, chemical inventors, etc. etc.; said it appeared to me that a Minister having such order at his disposal would find it very difficult to resist the pressure brought by such people as against the man of high science who

had not happened to have done anything to strike the
popular mind.

Discussed the possibility of submission of names by
somebody for the approval and choice of the Crown.   For
Science, I thought the R.S. Council might discharge that
duty very fairly.   I thought that the Academy of Berlin
presented people for the *Pour le Mérite*, but Lord S.
thought not.

In the course of conversation I spoke of Hooker's case
as a glaring example of the wrong way of treating distinguished men.   Observed that though I did not personally care for or desire the institution of such honorary
order, yet I thought it was a mistake in policy for the
Crown as the fountain of honour to fail in recognition
of that which deserves honour in the world of Science,
Letters, and Art.

Lord Salisbury smilingly summed up.   "Well, it
seems that you don't desire the establishment of such an
order, but that if you were in my place you would
establish it," to which I assented.

Said he had spoken to Leighton, who thought well of
the project.

It was not long, however, before he received im-
perative notice to quit town with all celerity.   He
fell ill with what turned out to be pleurisy ; and
after recruiting at Ilkley, went again to Switzerland.

4 MARLBOROUGH PLACE,
June 27, 1887.

MY DEAR FOSTER— . . . I am very sorry that it will
be impossible for me to attend [the meeting of committee
down for the following Wednesday].   If I am well
enough to leave the house I must go into the country
that day to attend the funeral of my wife's brother-in-
law and my very old friend Fanning, of whom I may

have spoken to you. He has been slowly sinking for some time, and this morning we had news of his death.

Things have been very crooked for me lately. I had a conglomerate of engagements of various degrees of importance in the latter half of last week, and had to forgo them all, by reason of a devil in the shape of muscular rheumatism of one side, which entered me last Wednesday, and refuses to be wholly exorcised (I believe it is my Jubilee Honour).[1] Along with it, and I suppose the cause of it, a regular liver upset. I am very seedy yet, and even if Fanning's death had not occurred I doubt if I should have been ready to face the Tyndall dinner.

The reference to this "Tyndall dinner" is explained in the following letters, which also refer to a meeting of the London University, in which the projects of reform which he himself supported met with a smart rebuff.

4 MARLBOROUGH PLACE,
May 13, 1887.

MY DEAR TYNDALL—I am very sorry to hear of your gout, but they say when it comes out at the toes it flies from the better parts, and that is to the good.

There is no sort of reason why unsatisfied curiosity should continue to disturb your domestic hearth ; your wife will have the gout too if it goes on. "They" can't bear the strain.

The history of the whole business is this. A day or two before I spoke to you, Lockyer told me that various

---

[1] On the same day he describes this to Sir J. Evans :—"I have hardly been out of the house as far as my garden, and not much off my bed or sofa since I saw you last. I have had an affection of the muscles of one side of my body, the proper name of which I do not know, but the similitude thereof is a bird of prey periodically digging in his claws and stopping your breath in a playful way."

people had been talking about the propriety of recognising your life-long work in some way or other; that, as you would not have anything else, a dinner had been suggested, and finally asked me to inquire whether you would accept that expression of goodwill. Of course I said I would, and I asked accordingly.

After you had assented I spoke to several of our friends who were at the Athenæum, and wrote to Lockyer. I believe a strong committee is forming, and that we shall have a scientific jubilation on a large scale; but I have purposely kept in the background, and confined myself, like Bismarck, to the business of " honest broker."

But of course nothing (beyond preliminaries) can be done till you name the day, and at this time of year it is needful to look well ahead if a big room is to be secured. So if you can possibly settle that point, pray do.

There seems to have been some oversight on my wife's part about the invitation, but she is stating her own case. We go on a visit to Mrs. Darwin to Cambridge on Saturday week, and the Saturday after that I am bound to be at Eton.

Moreover, I have sacrificed to the public Moloch so far as to promise to take the chair at a public meeting in favour of a Free Library for Marylebone on the 7th. As Wednesday's work at the Geological Society and the soirée knocked me up all yesterday, I shall be about finished I expect on the 8th. If you are going to be at Hindhead after that, and would have us for a day, it would be jolly; but I cannot be away long, as I have some work to finish before I go abroad.

I never was so uncomfortable in my life, I think, as on Wednesday when L—— was speaking, just in front of me, at the University. Of course I was in entire sympathy with the tenor of his speech, but I was no less certain of the impolicy of giving a chance to such a master of polished putting-down as the Chancellor. You know Mrs. Carlyle said that Owen's sweetness reminded her of sugar of lead. Granville's was that plus butter of antimony !—Ever yours very faithfully,

T. H. HUXLEY.

N.B.—Don't swear, but get Mrs. Tyndall, who is patient and good-tempered, to read this long screed.

May 18, 1887.

MY DEAR TYNDALL—I was very glad to get your letter yesterday morning, and I conveyed your alteration at once to Rücker, who is acting as secretary. I asked him to communicate with you directly to save time.

I hear that the proposal has been received very warmly by all sorts and conditions of men, and that is quite apart from any action of your closer personal friends. Personally I am rather of your mind about the " dozen or score " of the faithful. But as that was by no means to the mind of those who started the project, and, moreover, might have given rise to some heartburning, I have not thought it desirable to meddle with the process of spontaneous combustion. So look out for a big bonfire somewhere in the middle of June ! I have a hideous cold, and can only hope that the bracing air of Cambridge, where we go on Saturday, may set me right.— Ever yours very faithfully,   T. H. HUXLEY.

To recover from his pleuritic "Jubilee Honour" he went for a fortnight (July 11-25) to Ilkley, which had done him so much good before, intending to proceed to Switzerland as soon as he conveniently could.

ILKLEY, *July* 15, 1887.

MY DEAR FOSTER—I was very much fatigued by the journey here, but the move was good, and I am certainly mending, though not so fast as I could wish. I expect some adhesions are interfering with my bellows. As soon as I am fit to travel I am thinking of going to Lugano, and thence to Monte Generoso. The travelling is easy to Lugano, and I know the latter place.

My notion is I had better for the present avoid the chances of a wet, cold week in the high places.

M.B.A.[1] . . . As to the employment of the Grant, I think it ought to be on something definite and limited. The Pilchard question would be an excellent one to take up.

—— seems to have a notion of employing it on some geological survey of Plymouth Sound, work that would take years and years to do properly, and nothing in the way of clear result to show.

I hope to be in London on my way abroad in less than ten days' time, and will let you know.—Ever yours very faithfully,   T. H. HUXLEY.

And on the same day to Sir J. Donnelly :—

I expect . . . that I shall have a slow convalescence. Lucky it is no worse !

Much fighting I am likely to do for the Unionist cause or any other ! But don't take me for one of the enragés. If anybody will show me a way by which the Irish may attain all they want without playing the devil with us, I am ready to give them their own talking-shop or anything else.

But that is as much writing as I can sit up and do all at once.

---

[1] Marine Biological Association.

CHAPTER II

1887

ON the last day of July he left England for Switzerland, and did not return till the end of September. A second visit to Arolla worked a great change in him. He renewed his Gentian studies also, with unflagging ardour. The following letters give some idea of his doings and interests :—

HOTEL DU MONT COLLON, AROLLA, SWITZERLAND,
Aug. 28, 1887.

MY DEAR FOSTER—I know you will be glad to hear that I consider myself completely set up again. We went to the Maderaner Thal and stayed a week there. But I got no good out of it. It is charmingly pretty, but damp ; and, moreover, the hotel was 50 per cent too full of people, mainly Deutschers, and we had to turn out into the open air after dinner because the salon and fumoir were full of beds. So, in spite of all prudential considerations, I made up my mind to come here. We travelled over the Furca, and had a capital journey to Evolena. Thence I came on muleback (to my great disgust, but I could not walk a bit uphill) here. I began to get better at once ; and in spite of a heavy snowfall and arctic weather a week ago, I have done nothing but

29

mend. We have glorious weather now, and I can take almost as long walks as last year.

We have some Cambridge people here : Dr. Peile of Christ's and his family. Also Nettleship of Oxford. What is the myth about the Darwin tree in the Pall Mall ?[1] Dr. Peile believes it to be all a flam.

Forel has just been paying a visit to the Arolla glacier for the purpose of ascertaining the internal temperature. He told me he much desired to have a copy of the Report of the Krakatoa Committee. If it is published, will you have a copy sent to him ? He is Professor at Lausanne, and a very good man.

Our stay here will depend on the weather. At present it is perfect. I do not suppose we shall leave before 7th or 8th of September, and we shall get home by easy stages not much before the end of the month.—Ever yours very faithfully,   T. H. HUXLEY.

Madder than ever on Gentians.

The following is in reply to Sir E. Frankland's inquiries with reference to the reported presence of fish in the reservoirs of one of the water-companies.

HOTEL RIGHI VAUDOIS, GLION,
Sept. 16, 1887.

We left Arolla about ten days ago, and after staying a day at St. Maurice in consequence of my wife's indisposition, came on here where your letter just received has followed me. I am happy to say I am quite set up again, and as I can manage my 1500 or 2000 feet as well as ever, I may be pretty clear that my pleurisy has not left my lung sticking anywhere.

I will take your inquiries *seriatim.* (1) The faith of

---

[1] " A tree planted yesterday in the centre of the circular grass-plot in the first court of Christ's College, in Darwin's honour, was ' spirited ' away at night."—*P.M.G. August* 23, 1887.

your small boyhood is justified. Eels do wander overland, especially in the wet stormy nights they prefer for migration. But so far as I know this is the habit only of good-sized, downwardly-moving eels. I am not aware that the minute fry take to the land on their journey upwards.

(2) Male eels are now well known. I have gone over the evidence myself and examined many. But the reproductive organs of both sexes remain undeveloped in fresh water—just the contrary of salmon, in which they remain undeveloped in salt water.

(3) So far as I know, no eel with fully-developed reproductive organs has yet been seen. Their matrimonial operations go on in the sea where they spend their honeymoon, and we only know the result in the shape of the myriads of thread-like eel-lets which migrate up in the well-known " eel-fare."

(4) On general principles of eel-life I think it possible that the Inspector's theory *may* be correct. But your story about the roach is a poser. They certainly do not take to walking abroad. It reminds me of the story of the Irish milk-woman who was confronted with a stickleback found in the milk. " Sure, then, it must have been bad for the poor cow when that came through her teat."

Surely the Inspector cannot have overlooked such a crucial fact as the presence of other fish in the reservoirs ?

We shall be here another week, and then move slowly back to London. I am loth to leave this place, which is very beautiful with splendid air and charming walks in all directions—two or three thousand feet up if you like.

HOTEL RIGHI VAUDOIS, GLION, SWITZERLAND,
Sept. 16, 1887.

MY DEAR DONNELLY—We left Arolla for this place ten days ago, but my wife fell ill, and we had to stay a day at St. Maurice. She has been more or less out of sorts ever since until to-day. However, I hope now she is all right again.

This is a very charming place at the east end of the Lake of Geneva—1500 feet above the lake—and you can walk 3000 feet higher up if you like.

What they call a "funicular railway" hauls you up a gradient of 1 in $1\frac{3}{4}$ from the station on the shore in ten minutes. At first the sensation on looking down is queer, but you soon think nothing of it. The air is very fine, the weather lovely, the feeding unexceptionable, and the only drawback consists in the "javelins," as old Francis Head used to call them — stinks of such wonderful crusted flavour that they must have been many years in bottle. But this is a speciality of all furrin parts that I have ever visited.

I am very well and extremely lazy so far as my head goes—legs I am willing to use to any extent up hill or down dale. They wanted me to go and speechify at Keighley in the middle of October, but I could not get permission from the authorities. Moreover, I really mean to keep quiet and abstain even from good words (few or many) next session. My wife joins with me in love to Mrs. Donnelly.

She thought she had written, but doubts whether in the multitude of her letters she did not forget.—Ever yours, T. H. Huxley.

From Glion also he writes to Sir M. Foster :—

I have been doing some very good work on the Gentians in the interests of the business of being idle.

The same subject recurs in the next letter :—

Hotel Righi Vaudois, Glion, Switzerland,
Sept. 21, 1887.

My dear Hooker—I saw in the *Times* yesterday the announcement of Mr. Symonds' death. I suppose the deliverance from so painful a malady as heart-disease is hardly to be lamented in one sense ; but these increasing gaps in one's intimate circle are very saddening, and we feel for Lady Hooker and you. My wife has been greatly depressed by hearing of Mrs. Carpenter's fatal disorder. One cannot go away for a few weeks without finding somebody gone on one's return.

I got no good at the Maderaner Thal, so we migrated to our old quarters at Arolla, and there I picked up in no time, and in a fortnight could walk as well as ever. So if there are any adhesions they are pretty well stretched by this time.

I have been at the Gentians again, and worked out the development of the flower in *G. purpurea* and *G. campestris*. The results are very pretty. They both start from a thalamifloral condition, then become corollifloral, *G. purpurea* at first resembling *G. lutea* and *G. campestris*, an *Ophelia*, and then specialise to the *Ptychantha* and *Stephanantha* forms respectively.

In *G. campestris* there is another very curious thing. The anthers are at first introrse, but just before the bud opens they assume this position [sketch] and then turn right over and become extrorse. In *G. purpurea* this does not happen, but the anthers are made to open outwards by their union on the inner side of the slits of dehiscence.

There are several other curious bits of morphology have turned up, but I reserve them for our meeting.

Beyond pottering away at my Gentians and doing a little with that extraordinary *Cynanchum* I have been splendidly idle. After three weeks of the ascetic life of Arolla, we came here to acclimatise ourselves to lower levels and to fatten up. I go straight through the *table d'hôte* at each meal, and know not indigestion.

My wife has fared not so well, but she is all right again now. We go home by easy stages, and expect to be in Marlborough Place on Tuesday.

With all our best wishes to Lady Hooker and yourself—Ever yours,   T. H. HUXLEY.

The second visit to Arolla did as much good as
the first. Though unable to stay more than a week
or two in London itself, he was greatly invigorated.
His renewed strength enabled him to carry out
vigorously such work as he had put his hand to,
and still more, to endure one of the greatest sorrows
of his whole life which was to befall him this autumn
in the death of his daughter Marian.

The controversy which fell to his share immediately
upon his return, has already been mentioned (p. 11 *sq.*).
This was all part of the war for science which he
took as his necessary portion in life ; but he would
not plunge into any other forms of controversy,
however interesting. So he writes to his son, who
had conveyed him a message from the editor of a
political review :—

4 MARLBOROUGH PLACE,
Oct. 19, 1887.

No political article from me ! I have had to blow off
my indignation incidentally now and then lest worse
might befall me, but as to serious political controversy,
I have other fish to fry. Such influence as I possess
may be most usefully employed in promoting various
educational movements now afoot, and I do not want to
bar myself from working with men of all political parties.

So excuse me in the prettiest language at your
command to Mr. A.

Nevertheless politics very soon drew him into a
new conflict, in defence, be it said, of science against

the possible contamination of political influences. Prof. (now Sir) G. G. Stokes, his successor in the chair of the Royal Society, accepted an invitation from the University of Cambridge to stand for election as their member of Parliament, and was duly elected. This was a step to which many Fellows of the Royal Society, and Huxley in especial, objected very strongly. Properly to fulfil the duties of both offices at once was, in his opinion, impossible. It might seem for the moment an advantage that the accredited head of the scientific world should represent its interests officially in Parliament; but the precedent was full of danger. Science being essentially of no party, it was especially needful for such a representative of science to keep free from all possible entanglements; to avoid committing science, as it were, officially to the policy of a party, or, as its inevitable consequence, introducing political considerations into the choice of a future President.

During his own tenure of the Presidency Huxley had carefully abstained from any official connection with societies or public movements on which the feeling of the Royal Society was divided, lest as a body it might seem committed by the person and name of its President. He thought it a mistake that his successor should even be President of the Victoria Institute.

Thus there is a good deal in his correspondence bearing on this matter. He writes on November 6 to Sir J. Hooker :—

I am extremely exercised in my mind about Stokes'
going into Parliament (as a strong party man, moreover)
while still P.R.S.   I do not know what you may think
about it, but to my mind it is utterly wrong—and
degrading to the Society—by introducing politics into
its affairs.

And on the same day to Sir M. Foster :—

I think it is extremely improper for the President of
the R.S. to accept a position as a party politician.   As
a Unionist I should vote for him if I had a vote for
Cambridge University, but for all that I think it is most
lamentable that the Presidency of the Society should be
dragged into party mud.

When I was President I refused to take the Presidency
of the Sunday League, because of the division of opinion
on the subject.   Now we are being connected with the
Victoria Institute, and sucked into the slough of politics.

These considerations weighed heavily with several
both of the older and the younger members of the
Society ; but the majority were indifferent to the
dangers of the precedent.   The Council could not
discuss the matter ; they waited in vain for an official
announcement of his election from the President,
while he, as it turned out, expected them to broach
the subject.

Various proposals were discussed ; but it seemed
best that, as a preliminary to further action, an
editorial article written by Huxley should be inserted
in *Nature*, indicating what was felt by a section of
the Society, and suggesting that resignation of one
of the two offices was the right solution of the
difficulty.

Finally, it seemed that perhaps, after all, a "masterly inactivity" was the best line of action. Without risk of an authoritative decision of the Society "the wrong way," out of personal regard for the President, the question would be solved for him by actual experience of work in the House of Commons, where he would doubtless discover that he must "renounce either science, or politics, or existence."

This campaign, however, against a principle, was carried on without any personal feeling. The perfect simplicity of the President's attitude would have disarmed the hottest opponent, and indeed Huxley took occasion to write him the following letter, in reference to which he writes to Dr. Foster :—"I hate doing things in the dark and could not stand it any longer."

Dec. 1, 1887.

MY DEAR STOKES—When we met in the hall of the Athenæum on Monday evening I was on the point of speaking to you on a somewhat delicate topic ; namely, my responsibility for the leading article on the Presidency of R.S. and politics which appeared a fortnight ago in *Nature.* But I was restrained by the reflection that I had no right to say anything about the matter without the consent of the Editor of *Nature.* I have obtained that consent, and I take the earliest opportunity of availing myself of my freedom.

I should have greatly preferred to sign the article, and its anonymity is due to nothing but my strong desire to avoid the introduction of any personal irrelevancies into the discussion of a very grave question of principle.

I may add that as you are quite certain to vote in the way that I think right on the only political questions which greatly interest me, my action has not been, and cannot be, in any way affected by political feeling.

And as there is no one of whom I have a higher opinion as a man of science—no one whom I should be more glad to serve under, and to support year after year in the Chair of the Society, and no one for whom I entertain feelings of more sincere friendship—I trust you will believe that, if there is a word in the article which appears inconsistent with these feelings, it is there by oversight, and is sincerely regretted.

During the thirty odd years we have known one another, we have often had stout battles without loss of mutual kindness. My chief object in troubling you with this letter is to express the hope that, whatever happens, this state of things may continue. — I am, yours very faithfully,   T. H. HUXLEY.

P.S.—I am still of opinion that it is better that my authorship should not be officially recognised, but you are, of course, free to use the information I have given you in any way you may think fit.

To this the President returned a very frank and friendly reply ; saying he had never dreamed of any incompatibility existing between the two offices, and urging that the Presidency ought not to constrain a man to give up his ordinary duties as a citizen.   He concludes :—

And now I have stated my case as it appears to myself ; let me assure you that nothing that has passed tends at all to diminish my friendship towards you.   My wife heard last night that the article was yours, and told me so.   I rather thought it must have been written by some hot Gladstonian.   It seems, however, that her informant was right. She wishes me to tell you that she replied to her informant that she felt quite sure that if you wrote it, it was because you thought it.

To which Huxley replied :—

I am much obliged for your letter, which is just such as I felt sure you would write.

Pray thank Mrs. Stokes for her kind message. I am very grateful for her confidence in my uprightness of intention.

We must agree to differ.

It may be needful for me and those who agree with me to place our opinions on record ; but you may depend upon it that nothing will be done which can suggest any lack of friendship or respect for our President.

It will be seen from this correspondence and the letter to Sir J. Donnelly of July 15 (p. 28), that Huxley was a staunch Unionist. Not that he considered the actual course of English rule in Ireland ideal ; his main point was that under the circumstances the establishment of Home Rule was a distinct betrayal of trust, considering that on the strength of Government promises, an immense number of persons had entered into contracts, had bought land, and staked their fortunes in Ireland, who would be ruined by the establishment of Home Rule. Moreover, he held that the right of self-preservation entitled a nation to refuse to establish at its very gates a power which could, and perhaps would, be a danger to its own existence. Of the capacity of the Irish peasant for self-government he had no high opinion, and what he had seen of the country, and especially the great central plain, in his frequent visits to Ireland, convinced him that the balance between subsistence and population would speedily create a new agrarian question, whatever political schemes were introduced. This was one of "the only political questions which interested him."

Towards the end of October he left London for Hastings, partly for his own, but still more for his wife's sake, as she was far from well. He was still busy with one or two Royal Society Committees, and came up to town occasionally to attend their meetings, especially those dealing with the borings in the Delta, and with Antarctic exploration. Thus he writes :—

11 EVERSFIELD PLACE, HASTINGS,
Oct. 31, 1887.

MY DEAR FOSTER—We have been here for the last week, and are likely to be here for some time, as my wife, though mending, is getting on but slowly, and she will be as well out of London through beastly November. I shall be up on Thursday and return on Friday, but I do not want to be away longer, as it is lonesome for the wife.

I quite agree to what you propose on Committee, so I need not be there. Very glad to hear that the Council "very much applauded what we had done," and hope we shall get the £500.

I don't believe a word in increasing whale fishery, but scientifically, the Antarctic expedition would, or might be very interesting, and if the colonies will do their part, I think we ought to do ours.

You won't want me at that Committee either. Hope to see you on Thursday.—Ever yours,   T. H. HUXLEY.

Hideous pen !

But he did not come up that Thursday. His wife was for a time too ill to be left, and he winds up his letter of November 2 to Dr. Foster with the reflection :—

Man is born to trouble as the sparks, etc.—but when you have come to my time of life you will say as I do—Lucky it is no worse.

November 6.—I am very glad to hear that the £500 is granted, and I will see to what is next to be done as soon as I can. Also I am very glad to find you don't want my valuable service on Council R.S. I repented me of my offer when I thought how little I might be able to attend.

One thing, however, afforded him great pleasure at this time. He writes on November 6 to his old friend, Sir J. Hooker :—

I write just to say what infinite satisfaction the award of the Copley Medal to you has given me. If you were not my dear old friend, it would rejoice me as a mere matter of justice—of which there is none too much in this " —— rum world," as Whitworth's friend called it.

To the reply that the award was not according to rule, inasmuch as it was the turn for the medal to be awarded in another branch of science, he rejoins :—

I had forgotten all about the business—but he had done nothing to deserve the Copley, and all I can say is that if the present award is contrary to law, the "law's a hass" as Mr. Bumble said. But I don't believe that it is.

He replies also on November 5 to a clerical correspondent who had written to him on the distinc-

tion between *shehretz* and *rehmes*, and accused him of
" wilful blindness " in his theological controversy of
1886 :—

Let me assure you that it is not my way to set my
face against being convinced by evidence.

I really cannot hold myself to be responsible for the
translators of the Revised Version of the O.T.   If I had
given a translation of the passage to which you refer on
my own authority, any mistake would be mine, and I
should be bound to acknowledge it.   As I did not, I
have nothing to admit.   I have every respect for your
and Mr. ——'s authority as Hebraists, but I have noticed
that Hebrew scholars are apt to hold very divergent views,
and before admitting either your or Mr. ——'s interpretation, I should like to see the question fully discussed.

If, when the discussion is concluded, the balance of
authority is against the revised version, I will carefully
consider how far the needful alterations may affect the
substance of the one passage in my reply to Mr. Gladstone
which is affected by it.

At present I am by no means clear that it will make
much difference, and in no case · will the main lines of
my argument as to the antagonism between modern
science and the Pentateuch be affected.   The statements
I have made are public property.   If you think they are
in any way erroneous I must ask you to take upon yourself the same amount of responsibility as I have done,
and submit your objection to the same ordeal.

There is nothing like this test for reducing things to
their true proportions, and if you try it, you will probably
discover, not without some discomfort, that you really
had no reason to ascribe wilful blindness to those who do
not agree with you.

He was now preparing to complete his campaign
of the spring on technical education by delivering an

address to the Technical Education Association at Manchester on November 29, and looked forward to attending the anniversary meeting of the Royal Society on his way home next day, and seeing the Copley medal conferred upon his old friend, Sir J. Hooker. However, unexpected trouble befell him. First he was much alarmed about his wife, who had been ill more or less ever since leaving Arolla. Happily it turned out that there was nothing worse than could be set right by a slight operation. But nothing had been done when news came of the sudden death of his second daughter on November 19. "I have no heart for anything just now," he writes; nevertheless, he forced himself to fulfil this important engagement at Manchester, and in the end the necessity of bracing himself for the undertaking acted on him as a tonic.

It is a trifle, perhaps, but a trifle significant of the disturbance of mind that could override so firmly fixed a habit, that the two first letters he wrote after receiving the news are undated; almost the only omission of the sort I have found in all his letters of the last twenty-five years of his life.

His daughter's long illness had left him without hope for months past, but this, as he confessed, did not mend matters much. In his letters to his two most intimate friends, he recalls her brilliant promise, her happy marriage, her "faculty for art, which some of the best artists have told me amounted to genius." But he was naturally reticent in these matters, and would hardly write of his own griefs unbidden even to old friends.

85 MARINA, ST. LEONARDS,
Nov. 21, 1887.

MY DEAR SPENCER—You will not have forgotten my bright girl Marian, who married so happily and with such bright prospects half a dozen years ago ?

Well, she died three days ago of a sudden attack of pneumonia, which carried her off almost without warning. And I cannot convey to you a sense of the terrible sufferings of the last three years better than by saying that I, her father, who loved her well, am glad that the end has come thus. . . .

My poor wife is well nigh crushed by the blow. For though I had lost hope, it was not in the nature of things that she should.

Don't answer this—I have half a mind to tear it up —for when one is in a pool of trouble there is no sort of good in splashing other people.—Ever yours,

T. H. HUXLEY.

As for his plans, he writes to Sir J. Hooker on November 21 :—

I had set my heart on seeing you get the Copley on the 30th. In fact, I made the Manchester people, to whom I had made a promise to go down and address the Technical Education Association, change their day to the 29th for that reason.

I cannot leave them in the lurch after stirring up the business in the way I have done, and I must go and give my address. But I must get back to my poor wife as fast as I can, and I cannot face any more publicity than that which it would be cowardly to shirk just now. So I shall not be at the Society except in the spirit.—Ever yours,   T. H. HUXLEY.

And again to Sir M. Foster :—

You cannot be more sorry than I am that I am going to Manchester, but I am not proud of chalking up "no popery" and running away—for all Evans' and your chaff—and, having done a good deal to stir up the Technical Education business and the formation of the Association, I cannot leave them in the lurch when they urgently ask for my services. . . .

The Delta business must wait till after the 30th. I have no heart for anything just now.

The letters following were written in answer to letters of sympathy.

85 MARINA, ST. LEONARDS,
Nov. 25, 1887.

MY DEAR MR. CLODD—Let me thank you on my wife's behalf and my own for your very kind and sympathetic letter.

My poor child's death is the end of more than three years of suffering on her part, and deep anxiety on ours. I suppose we ought to rejoice that the end has come, on the whole, so mercifully. But I find that even I, who knew better, hoped against hope, and my poor wife, who was unfortunately already very ill, is quite heart-broken. Otherwise, she would have replied herself to your very kind letter.

She has never yet learned the art of sparing herself, and I find it hard work to teach her.—Ever yours very faithfully, T. H. HUXLEY.

In the same strain he writes to Dr. Dyster :—

Rationally we must admit that it is best so. But then, whatever Linnæus may say, man is not a rational animal —especially in his parental capacity.

85 MARINA, ST. LEONARDS,
Nov. 25, 1887.

MY DEAR KNOWLES—I really must thank you very heartily for your letter. It went to our hearts and did us good, and I know you will like to learn that you have helped us in this grievous time.

My wife is better, but fit for very little; and I do not let her write a letter even, if I can help it. But it is a great deal harder to keep her from doing what she thinks her duty than to get most other people to do what plainly is their duty.

With our kindest love and thanks to all of you—Ever, my dear Knowles, yours very faithfully,   T. H. HUXLEY.

Yes, you are quite right about "loyal." I love my friends and hate my enemies, which may not be in accordance with the Gospel, but I have found it a good wearing creed for honest men.

The "Address on behalf of the National Association for the Promotion of Technical Education," first published in the ensuing number of *Science and Art*, and reprinted in *Collected Essays*, iii. 427-451, was duly delivered in Manchester, and produced a considerable effect.

He writes to Sir M. Foster, December 1:—

I am glad I resisted the strong temptation to shirk the business. Manchester has gone solid for technical education, and if the idiotic London papers, instead of giving half a dozen lines of my speech, had mentioned the solid contributions to the work announced at the meeting, they would have enabled you to understand its importance.

. . . I have the satisfaction of having got through a hard bit of work, and am none the worse physically—rather the better for having to pull myself together.

And to Sir J. Hooker :—

> 85 Marina, St. Leonards,
> *Dec.* 4, 1887.

My dear Hooker—$x = 8$, 6.30. I meant to have written to ask you all to put off the x till next Thursday, when I could attend, but I have been so bedevilled I forgot it. I shall ask for a bill of indemnity.

I was rather used up yesterday, but am picking up. In fact my Manchester journey convinced me that there was more stuff left than I thought for. I travelled 400 miles, and made a speech of fifty minutes in a hot, crowded room, all in about twelve hours, and was none the worse. Manchester, Liverpool, and Newcastle have now gone in for technical education on a grand scale, and the work is practically done. *Nunc Dimittis!*

I hear great things of your speech at the dinner. I wish I could have been there to hear it. . . .

Of the two following letters, one refers to the account of Sir J. D. Hooker's work in connection with the award of the Copley medal ; the other, to Hooker himself, touches a botanical problem in which Huxley was interested.

> St. Leonards, *Nov.* 25, 1887.

My dear Foster— . . . I forget whether in the notice of Hooker's work you showed me there was any allusion made to that remarkable account of the Diatoms in Antarctic ice, to which I once drew special attention, but Heaven knows where ?

Dyer perhaps may recollect all about the account in the *Flora Antarctica,* if I mistake not. I have always looked upon Hooker's insight into the importance of these things and their skeletons as a remarkable piece of inquiry —anticipative of subsequent deep sea work.

Best thanks for taking so much trouble about H——.
Pray tell him if ever you write that I have not answered
his letter only because I awaited your reply. He may
think my silence uncivil. . . .—Ever yours,

T. H. HUXLEY.

TO SIR J. D. HOOKER

4 MARLBOROUGH PLACE,
Dec. 29, 1887.

Where is the fullest information about distribution of
Coniferæ ? Of course I have looked at *Genera Pl.* and
De Candolle.

I have been trying to make out whether structure or
climate or paleontology throw any light on their distribution—and am drawing complete blank. Why the deuce
are there no Conifers but *Podocarpus* and *Widringtonias*
in all Africa south of the Sahara ? And why the double
deuce are about three-quarters of the genera huddled
together in Japan and N. China ?

I am puzzling over this group because the paleontological record is comparatively so good.

I am beginning to suspect that present distribution is
an affair rather of denudation than migration.

Sequoia ! Taxodium ! Widringtonia ! Araucaria ! all
in Europe, in Mesozoic and Tertiary.

The following letters to Mr. Herbert Spencer were
written as sets of proofs of his Autobiography arrived.
That to Sir J. Skelton was to thank him for his book
on *Maitland of Lethington,* the Scotch statesman of
the time of Queen Mary.

Jan. 18, 1887.

(The first part of this letter is given on p. 3.)

MY DEAR SPENCER—I see that your proofs have been
in my hands longer than I thought for. But you may

have seen that I have been "starring" at the Mansion
House. . . .

I am immensely tickled with your review of your
own book. That is something most originally Spencerian.
I have hardly any suggestions to make, except in what
you say about the *Rattlesnake* work and my position on
board.

Her proper business was the survey of the so-called
"inner passage" between the Barrier Reef and the east
coast of Australia; the New Guinea work was a *hors
d'œuvre*, and dealt with only a small part of the southern
coast.

Macgillivray was naturalist—I was actually Assistant-
Surgeon and nothing else. But I was recommended to
Stanley by Sir John Richardson, my senior officer at
Haslar, on account of my scientific proclivities. But
scientific work was no part of my duty. How odd it is
to look back through the vista of years! Reading your
account of me, I had the sensation of studying a fly in
amber. I had utterly forgotten the particular circumstance that brought us together. Considering what wilful
tykes we both are (you particularly), I think it is a great
credit to both of us that we are firmer friends now than
we were then. Your kindly words have given me much
pleasure.

This is a deuce of a long letter to inflict upon you,
but there is more coming. The other day a Miss ——,
a very good, busy woman of whom I and my wife have
known a little for some years, sent me a proposal of the
committee of a body calling itself the London Liberty
League (I think) that I should accept the position of one
of three honorary something or others, you and Mrs.
Fawcett being the other two.

Now you may be sure that I should be glad enough
to be associated with you in anything; but considering
the innumerable battles we have fought over education,
vaccination, and so on, it seemed to me that if the pro-

gramme of the League were wide enough to take us both for figure-heads, it must be so elastic as to verge upon infinite extensibility ; and that one or other of us would be in a false position.

So I wrote to Miss —— to that effect, and the matter then dropped.

Misrepresentation is so rife in this world that it struck me I had better tell you exactly what happened.

On the whole, your account of your own condition is encouraging ; not going back is next door to going forward. Anyhow, you have contrived to do a lot of writing.

We are all pretty flourishing, and if my wife does not get worn out with cooks falling ill and other domestic worries, I shall be content.

Now this really is the end.—Ever yours very truly,
T. H. HUXLEY.

4 MARLBOROUGH PLACE, LONDON, N.W.,
March 7, 1887.

MY DEAR SKELTON—Wretch that I am, I see that I have never had the grace to thank you for *Maitland of Lethington* which reached me I do not choose to remember how long ago, and which I read straight off with lively satisfaction.

There is a paragraph in your preface, which I meant to have charged you with having plagiarised from an article of mine, which had not appeared when I got your book. In that Hermitage of yours you are up to any Esotericobuddhistotelepathic dodge !

It is about the value of practical discipline to historians. Half of them know nothing of life, and still less of government and the ways of men.

I am quite useless, but have vitality enough to kick and scratch a little when prodded.

I am at present engaged on a series of experiments on the thickness of skin of that wonderful little wind-bag ——. The way that second-rate amateur poses as a man of science, having authority as a sort of papistical Scotch dominie, bred a minister, but stickit, really "rouses my corruption." What a good phrase that is. I am cursed with a lot of it, and any fool can strike ile. —Ever yours very faithfully,   T. H. Huxley.[1]

Please remember me very kindly to Mrs. Skelton.

> 11 Eversfield Place, Hastings,
> *Nov.* 18, 1887.

My dear Spencer—I was very glad to get your letter this morning. I heard all about you from Hirst before I left London, now nearly a month ago, and I promised myself that instead of bothering you with a letter I would run over from here and pay you a visit.

Unfortunately, my wife, who had been ill more or less ever since we left Arolla and came here on Clark's advice, had an attack one night, which frightened me a good deal, though it luckily turned out to arise from easily remediable causes.

Under these circumstances you will understand how I have not made my proposed journey to Brighton.

I am rejoiced to hear of your move. I believe in the skill of Dr. B. Potter and her understanding of the case more than I do in all the doctors and yourself put together. Please offer my respectful homage to that eminent practitioner.

You see people won't let me alone, and I have had to tell the Duke to "keep on board his own ship," as the Quaker said, once more. I seek peace, but do not ensue it.

Send any quantity of proofs, they are a good sign. By the way, we move to 85 Marina, St. Leonards, to-morrow.

Wife sends her kind regards. — Ever yours very faithfully,   T. H. Huxley.

---

[1] This letter is one of the twelve from T. H. H. already published by Sir John Skelton in his *Table Talk of Shirley*, p. 295 *sq.*

85 Marina, St. Leonards,
Dec. 1887.

My dear Spencer—I have nothing to criticise in the
enclosed except that the itineraries seem to me rather
superfluous.

I am glad to find that you forget things that have
happened to you as completely as I do. I should cut
almost as bad a figure as "Sir Roger" if I were cross-
examined about my past life.

Your allusion to sending me the proofs made me laugh
by reminding me of a particularly insolent criticism with
which I once favoured you : " No objection except to the
whole."

It was some piece of diabolical dialectics, in which I
could pick no hole, if the premises were granted—and
even then could be questioned only by an ultra-sceptic !

Do you see that the American Association of Authors
has adopted a Resolution, which is a complete endorsement
of my view of the stamp-swindle ?

We have got our operation over, and my wife is going
on very well.   Overmuch anxiety has been telling on me,
but I shall throw it off.—Ever yours very faithfully,

T. H. Huxley.

CHAPTER III

1888

HUXLEY had returned to town before Christmas, for the house in St. John's Wood was still the rallying-point for the family, although his elder children were now married and dispersed. But he did not stay long. "Wife wonderfully better," he writes to Sir M. Foster on January 8, "self as melancholy as a pelican in the wilderness." He meant to have left London on the 16th, but his depressed condition proved to be the beginning of a second attack of pleurisy, and he was unable to start for Bournemouth till the 24th.

Here, however, his recovery was very slow. He was unable to come up to the first meeting of the x Club. "I trust," he writes, "I shall be able to be at the next x—but I am getting on very slowly. I can't walk above a couple of miles without being exhausted, and talking for twenty minutes has the same effect. I suppose it is all Anno Domini."

But he had a pleasant visit from one of the x, and writes :—

CASALINI, WEST CLIFF, BOURNEMOUTH,
Jan. 29, 1888.

MY DEAR HOOKER—Spencer was here an hour ago as lively as a cricket. He is going back to town on Tuesday to plunge into the dissipations of the Metropolis. I expect he will insist on you all going to Evans' (or whatever represents that place to our descendants) after the *x*.

Bellows very creaky—took me six weeks to get them mended last time, so I suppose I may expect as long now. —Ever yours very faithfully,   T. H. HUXLEY.

As appears from the letters which follow, he had been busied with writing an article for the *Nineteenth Century*, for February, on the "Struggle for Existence," [1] which on the one hand ran counter to some of Mr. Herbert Spencer's theories of society ; and on the other, is noticeable as briefly enunciating the main thesis of his "Romanes Lecture" of 1893.

85 MARINA, ST. LEONARDS,
Dec. 13, 1887.

MY DEAR KNOWLES—I have to go to town to-morrow for a day, so that puts an end to the possibility of getting my screed ready for January. Altogether it will be better to let it stand over.

I do not know whence the copyright extract came, except that, as Putnam's name was on the envelope, I suppose they sent it.

Pearsall Smith's practice is a wonderful commentary on his theory. Distribute the contents of the baker's shop *gratis*—it will give people a taste for bread !

Great is humbug, and it will prevail, unless the

---

[1] *Coll. Ess.* ix. 195.

people who do not like it hit hard.   The beast has no brains, but you can knock the heart out of him.—Ever yours very truly,   T. H. HUXLEY.

> 4 MARLBOROUGH PLACE,
> *Jan.* 9, 1888.

MY DEAR DONNELLY—Here is my proof.   Will you mind running your eye over it?

The article is long, and partly for that reason and partly because the general public wants principles rather than details, I have condensed the practical half.

H. Spencer and "Jus" will be in a white rage with me.—Ever yours very faithfully,   T. H. HUXLEY.

To Professor Frankland, February 6 :—

I am glad you like my article.   There is no doubt it is rather like a tadpole, with a very big head and a rather thin tail.   But the subject is a ticklish one to deal with, and I deliberately left a good deal suggested rather than expressed.

> CASALINI, WEST CLIFF, BOURNEMOUTH,
> *Feb.* 9, 1888.

MY DEAR DONNELLY—No! I don't think softening has begun yet—*vide* "Nature" this week.[1]   I am glad you found the article worth a second go.   I took a vast of trouble (as the country folks say) about it.   I am afraid it has made Spencer very angry—but he knows I think he has been doing mischief this long time.

Bellows to mend!   Bellows to mend!   I am getting very tired of it.   If I walk two or three miles, however

---

[1] *Nature* (xxxvii. 337) for February 9, 1888 : review of his article in the *Nineteenth Century* on the "Industrial Struggle for Existence."

slowly, I am regularly done for at the end of it.   I
expect there has been more mischief than I thought for.

How about the Bill ?—Ever yours,

T. H. HUXLEY.

However, he and Mr. Spencer wrote their minds
to each other on the subject, and as Huxley remarks
with reference to this occasion, "the process does us
both good, and in no way interferes with our
friendship."

The letter immediately following, to Mr. Romanes,
answers an inquiry about a passage quoted from
Huxley's writings by Professor Schurman in his
Ethical Import of Darwinism.   This passage, made up
of sentences from two different essays, runs as
follows :—

It is quite conceivable that every species tends to
produce varieties of a limited number and kind, and that
the effect of natural selection is to favour the development of some of these, while it opposes the development
of others along their predetermined line of modification.[1]
A whale does not tend to vary in the direction of producing feathers, nor a bird in the direction of producing
whalebone.[2]

"On the strength of these extracts" (writes Mr.
Romanes), "Schurman represents you 'to presuppose
design, since development takes place along certain predetermined lines of modification.' But as he does not
give references, and as I do not remember the passages, I
cannot consult the context, which I fancy must give a
different colouring to the extracts."

---

[1] *Coll. Ess.* ii. 223.
[2] In "Mr. Darwin's Critics," 1871 ; *Coll. Ess.* ii. 181.

4 MARLBOROUGH PLACE,
Jan. 5, 1888.

MY DEAR ROMANES—They say that liars ought to have long memories. I am sure authors ought to. I could not at first remember where the passage Schurman quotes occurs, but I did find it in the *Encyclopædia Britannica* article on "Evolution,"[1] reprinted in *Science and Culture*, p. 307.

But I do not find anything about the "whale" here. Nevertheless I have a consciousness of having said something of the kind somewhere.[2]

If you look at the whole passage, you will see that there is not the least intention on my part to presuppose design.

If you break a piece of Iceland spar with a hammer, all the pieces will have shapes of a certain kind, but that does not imply that the Iceland spar was constructed for the purpose of breaking up in this way when struck. The atomic theory implies that of all possible compounds of A and B only those will actually exist in which the proportions of A and B by weight bear a certain numerical ratio. But it is mere arguing in a circle to say that the fact being so is evidence that it was designed to be so.

I am not going to take any more notice of the everlasting D——, as you appropriately call him, until he has withdrawn his slanders. . . .

Pray give him a dressing—it will be one of those rare combinations of duty and pleasure.—Ever yours very faithfully,   T. H. HUXLEY.

He was, moreover, constantly interested in schemes for the reform of the scientific work of the London University, and for the enlargement of the

---

[1] *Coll. Ess.* ii. 223.
[2] In "Mr. Darwin's Critics," 1871; *Coll. Ess.* ii. 181.

scope and usefulness of the Royal Society. As for
the latter, a proposal had been made for federation
with colonial scientific societies, which was opposed
by some of his friends in the *x* Club; and he writes
to Sir E. Frankland on February 3 :—

> I am very sorry you are all against Evans' scheme. I
> am for it. I think it a very good proposal, and after all
> the talk, I do not want to see the Society look foolish by
> doing nothing.
>
> You are a lot of obstructive old Tories, and want
> routing out. If I were only younger and less indisposed
> to any sort of exertion, I would rout you out finely !

With respect to the former, it had been proposed
that medical degrees should be conferred, not by the
university, but by a union of the several colleges
concerned. He writes :—

> 4 MARLBOROUGH PLACE,
> *Jan.* 11, 1888.
>
> MY DEAR FOSTER—I send back the "Heathen
> Deutscheree's" (whose ways are dark) letter lest I forget
> it to-morrow.
>
> Meanwhile perpend these two things :—
>
> 1. United Colleges propose to give just as good an
> examination and require as much qualification as the
> Scotch Universities. Why then give their degree a
> distinguishing mark ?
>
> 2. "Academical distinctions" in medicine are all
> humbug. You are making a medical technical school at
> Cambridge—and quite right too. The United Colleges,
> if they do their business properly, will confer just as
> much, or as little "academical distinction" as Cambridge
> by their degree.
>
> 3. The Fellowship of the College of Surgeons is in

every sense as much an "academical distinction" as the Masterships in Surgery or Doctorate of Medicine of the Scotch and English Universities.

4. You may as well cry for the moon as ask my colleagues in the Senate to meddle seriously with the Matriculation. They are possessed by the devil that cries continually, "There is only the Liberal education, and Greek and Latin are his prophets."

At Bournemouth he also applied himself to writing the Darwin obituary notice for the Royal Society, a labour of love which he had long felt unequal to undertaking. The MS. was finally sent off to the printer's on April 6, unlike the still longer unfinished memoir on *Spirula*, to which allusion is made here, among other business of the *Challenger* Committee, of which he was a member.

On February 12 he writes to Sir J. Evans:—

Spirula is a horrid burden on my conscience—but nobody could make head or tail of the business but myself.

That and Darwin's obituary are the chief subjects of my meditations when I wake in the night. But I do not get much "forrarder," and I am afraid I shall not until I get back to London.

BOURNEMOUTH, *Feb.* 14, 1888.

MY DEAR FOSTER—No doubt the Treasury will jump at any proposition which relieves them from further expense—but I cannot say I like the notion of leaving some of the most important results of the *Challenger* voyage to be published elsewhere than in the official record. . . .

Evans made a deft allusion to *Spirula*, like a powder

between two dabs of jam.   At present I have no moral sense, but it may awake as the days get longer.

I have been reading the *Origin* slowly again for the *n*th time, with the view of picking out the essentials of the argument, for the obituary notice.   Nothing entertains me more than to hear people call it easy reading.

Exposition was not Darwin's *forte*—and his English is sometimes wonderful.   But there is a marvellous dumb. sagacity about him—like that of a sort of miraculous dog—and he gets to the truth by ways as dark as those of the Heathen Chinee.

I am getting quite sick of all the "paper philosophers," as old Galileo called them, who are trying to stand upon Darwin's shoulders and look bigger than he, when in point of real knowledge they are not fit to black his shoes.   It is just as well I am collapsed or I believe I should break out with a final " Für Darwin."

I will think of you when I get as far as the fossils. At present I am poking over *P. sylvestris* and *P. pinnata* in the intervals of weariness.

My wife joins with me in love to you both.—Ever yours very faithfully,   T. H. HUXLEY.

Snow and cold winds here.   Hope you are as badly off at Cambridge.

BOURNEMOUTH, *Feb.* 21, 1888.

MY DEAR FOSTER—We have had nothing but frost and snow here lately, and at present half a gale of the bitterest north-easter I have felt since we were at Florence is raging.[1]

I believe I am getting better, as I have noticed that at a particular stage of my convalescence from any sort of illness I pass through a condition in which things in

---

[1] Similarly to Sir J. Evans on the 28th—" I get my strength back but slowly, and think of migrating to Greenland or Spitzbergen for a milder climate."

general appear damnable and I myself an entire failure.
If that is a sign of returning health you may look upon
my restoration as certain.

If it is only Murray's speculations he wants to publish
separately, I should say by all means let him. But the
facts, whether advanced by him or other people, ought
all to be in the official record. I agree we can't stir.

I scented the "goak." How confoundedly proud you
are of it. In former days I have been known to joke
myself.

I will look after the questions if you like. In my
present state of mind I shall be a capital critic—on
Dizzy's views of critics. . . .—Ever yours,   T. H. H.

This year Huxley was appointed a Trustee of the
British Museum, an office which he had held *ex officio*
from 1883 to 1885, as President of the Royal
Society.

This is referred to in the following letter of
March 9 :—

MY DEAR HOOKER—Having nothing to do plays the
devil with doing anything, and I suppose that is why I
have been so long about answering your letter.

There is nothing the matter with me now except want
of strength. I am tired out with a three-mile walk, and
my voice goes if I talk for any time. I do not suppose
I shall do much good till I get into high and dry air,
and it is too early for Switzerland yet. . . .

You see I was honoured and gloried by a trusteeship
of the B.M.[1] These things, I suppose, normally come

---

[1] Replying on the 2nd to Sir John Evans' congratulations, he
says :—" It is some months since Lord Salisbury made the proposal
to me, and I was beginning to wonder what had happened—
whether Cantuar had put his foot down for example, and objected
to bad company."

when one is worn-out. When Lowe was Chancellor of the Exchequer I had a long talk with him about the affairs of the Nat. Hist. Museum, and I told him that he had better put Flower at the head of it and make me a trustee to back him. Bobby no doubt thought the suggestion cheeky, but it is odd that the thing has come about now that I don't care for it, and desire nothing better than to be out of every description of bother and responsibility.

Have not Lady Hooker and you yet learned that a large country house is of all places the most detestable in cold weather? The neuralgia was a mild and kindly hint of Providence not to do it again, but I am rejoiced it has vanished.

Pronouns got mixed somehow.

With our kindest regards—Ever yours,

T. H. HUXLEY.

More last words :—What little faculty I have has been bestowed on the obituary of Darwin for R.S. lately. I have been trying to make it an account of his intellectual progress, and I hope it will have some interest. Among other things I have been trying to set out the argument of the "Origin of Species," and reading the book for the *n*th time for that purpose. It is one of the hardest books to understand thoroughly that I know of, and I suppose that is the reason why even people like Romanes get so hopelessly wrong.

If you don't mind, I should be glad if you would run your eye over the thing when I get as far as the proof stage—Lord knows when that will be.

A few days later he wrote again on the same subject, after reading the obituary of Asa Gray, the first American supporter of Darwin's theory.

March 23.—I suppose Dana has sent you his obituary of Asa Gray.

The most curious feature I note in it is that neither of them seems to have mastered the principles of Darwin's theory.   See the bottom of p. 19 and the top of p. 20. As I understand Darwin there is nothing "Anti-Darwinian" in either of the two doctrines mentioned.

Darwin has left the causes of variation and the question whether it is limited or directed by external conditions perfectly open.

The only serious work I have been attempting lately is Darwin's obituary.   I do a little every day, but get on very slowly.   I have read the life and letters all through again, and the *Origin* for the sixth or seventh time, becoming confirmed in my opinion that it is one of the most difficult books to exhaust that ever was written.

I have a notion of writing out the argument of the *Origin* in systematic shape as a sort of primer of *Darwinismus*.   I have not much stuff left in me, and it would be as good a way of using what there is as I know of.   What do you think ?—Ever yours,

T. H. HUXLEY.

In reply to this Sir J. Hooker was inclined to make the biographer alone responsible for the confusion noted in the obituary of Asa Gray.   He writes :—

March 27, 1888.

DEAR HUXLEY—Dana's Gray arrived yesterday, and I turned to pp. 19, 20.   I see nothing Anti-Darwinian in the passages, and I do not gather from them that Gray did.

I did not follow Gray into his later comments on Darwinism, and I never read his *Darwiniana*.   My recollection of his attitude after acceptance of the doctrine, and during the first few years of his active promulgation of it, is that he understood it clearly, but sought to harmonise it with his prepossessions, without disturbing its physical principles in any way.

He certainly showed far more knowledge and appreciation of the contents of the *Origin* than any of the reviewers and than any of the commentators, yourself excepted.

Latterly he got deeper and deeper into theological and metaphysical wanderings, and finally formulated his ideas in an illogical fashion.

. . . Be all this as it may, Dana seems to be in a muddle on p. 20, and quite a self-sought one.—Ever yours,   J. D. HOOKER.

The following is a letter of thanks to Mrs. Humphry Ward for her novel *Robert Elsmere.*

BOURNEMOUTH, *March* 15, 1888.

MY DEAR MRS. WARD—My wife thanked you for your book which you were so kind as to send us. But that was grace before meat, which lacks the " physical basis " of after-thanksgiving—and I am going to supplement it, after my most excellent repast.

I am not going to praise the charming style, because that was in the blood and you deserve no sort of credit for it. Besides, I should be stepping beyond my last. But as an observer of the human ant-hill—quite impartial by this time—I think your picture of one of the deeper aspects of our troubled time admirable.

You are very hard on the philosophers. I do not know whether Langham or the Squire is the more unpleasant—but I have a great deal of sympathy with the latter, so I hope he is not the worst.

If I may say so, I think the picture of Catherine is the gem of the book. She reminds me of her namesake of Siena— and would as little have failed in any duty, however gruesome. You remember Sodoma's picture.

Once more, many thanks for a great pleasure.
My wife sends her love.—Ever yours very faithfully,

T. H. HUXLEY.

Meanwhile, he had been making no progress towards health; indeed, was going slowly downhill. He makes fun of his condition when writing to condole with Mr. Spencer on falling ill again after the unwonted spell of activity already mentioned; but a few weeks later discovered the cause of his weakness and depression in an affection of the heart. This was not immediately dangerous, though he looked a complete wreck. His letters from April onwards show how he was forced to give up almost every form of occupation, and even to postpone his visit to Switzerland, until he had been patched up enough to bear the journey.

> CASALINI, WEST CLIFF, BOURNEMOUTH,
> *March* 9, 1888.

MY DEAR SPENCER—I am very sorry to hear from Hooker that you have been unwell again. You see if young men from the country will go plunging into the dissipations of the metropolis nemesis follows.

Until two days ago, the weathercocks never overstepped N. on the one side and E. on the other ever since you left. Then they went west with sunshine and most enjoyable softness—but next S. with a gale and rain—all ablowin' and agrowin' at this present.

I have nothing to complain of so long as I do nothing; but although my hair has grown with its usual rapidity I differ from Samson in the absence of a concurrent return of strength. Perhaps that is because a male hairdresser, and no Delilah, cut it last! But I waste Biblical allusions upon you.

My wife and Nettie, who is on a visit, join with me in best wishes.

Please let me have a line to say how you are—Gladstonianly on a post-card.—Ever yours very faithfully,

T. H. HUXLEY.

BOURNEMOUTH, *April 7*, 1888.

MY DEAR FOSTER—"Let thy servant's face be white before thee." The obituary of Darwin went to Rix [1] yesterday! It is not for lack of painstaking if it is not worth much, but I have been in a bad vein for work of any kind, and I thought I should never get even this simple matter ended.

I have been bothered with præcordial uneasiness and intermittent pulse ever since I have been here, and at last I got tired of it and went home the day before yesterday to get carefully overhauled. Hames tells me there is weakness and some enlargement of the left ventricle, which is pretty much what I expected. Luckily the valves are all right.

I am to go and devote myself to coaxing the left v. wall to thicken *pro rata*—among the mountains, and to have nothing to do with any public functions or other exciting bedevilments. So the International Geological Congress will not have the pleasure of seeing its Honorary President in September. I am disgusted at having to break an engagement, but I cannot deny that Hames is right. At present the mere notion of the thing puts me in a funk.

I wish I could get out of the chair of the M.B.A. also. . . . I know that you and Evans and Dyer will do your best, but you are all eaten up with other occupations.

---

[1] Assistant Secretary of the Royal Society.

Just turn it over in your mind—there's a dear good fellow—just as if you hadn't any other occupations.

With which eminently reasonable and unselfish request believe me—Ever yours,   T. H. H.

BOURNEMOUTH, April 10, 1888.

MY DEAR FOSTER—I send by this post the last—I hope for your sake and for that of the recording angel— of ——.[1] I agree to all Brady's suggestions.

With all our tinkering I feel inclined to wind up the affair after the manner of Mr. Shandy's summing [up] of the discussion about Tristram's breeches—"And when he has got 'em he'll look a beast in 'em."—Ever yours,
T. H. H.

April 12.   To the same :—

I am quite willing to remain at the M.B.A. till the opening.   If Evans will be President I shall be happy.

—— is a very good man, but you must not expect too much of the "wild-cat" element, which is so useful in the world, in him.

I am disgusted with myself for letting everything go by the run, but there is no help for it.   The least thing bowls me over just now.

CASALINI, WEST CLIFF, BOURNEMOUTH,
April 12, 1888.

MY DEAR HOOKER—I plead not guilty.[2]   It was agreed at the last meeting that there should be none in April—I suppose by reason of Easter, so I sent no notice. This is what Frankland told me in his letter of the 2nd. However, I see you were present, so I can't make it out.

---

[1] The "Heathen Deutscheree" of p. 58.   A paper of his, contributed to the Royal Society, had been under revision.

[2] In the matter of sending out no notices for a meeting of the *x* Club.

My continual absence makes me a shocking bad
Treasurer, and I am sorry to say that things will be
worse instead of better. Ever since this last pleuritic
business I have been troubled with præcordial uneasiness.
[After an account of his symptoms he continues] So I am
off (with my wife) to Switzerland at the end of this
month, and shall be away all the summer. We have not
seen the Engadine and Tyrol yet, so we shall probably
make a long circuit. It is a horrid nuisance to be exiled
in this fashion. I have hardly been at home one month
in the last ten. But it is of no use to growl.

Under these circumstances, would you mind looking
after the x while I am away? There is nothing to do but
to send the notices on Saturday previous to the meeting.

I am very grieved to hear about Hirst—though to say
truth, the way he has held out for so long has been a
marvel to me. The last news I had of Spencer was not
satisfactory.

Eheu! the "Table Round" is breaking up. It's a
great pity; we were such pleasant fellows, weren't we?—
Ever yours,   T. H. HUXLEY.

CASALINI, WEST CLIFF, BOURNEMOUTH,
April 18, 1888.

MY DEAR FOSTER—I am cheered by your liking of the
notice of Darwin. I read the "Life and Letters," and
the "Origin," Krause's "Life," and some other things over
again in order to do it. But I have not much go in me,
and I was a scandalous long time pottering over the
writing.

I have sent the proof back with a variety of interpola-
tions. I would have brought the "Spirula" notes down
here to see what I could do, but I felt pretty sure that if
I brought two things I should not do one. Nobody could
do anything with it but myself. I will try what I can
do when I go to town. How much time is there before
the wind-up of the Challenger?

We go up to town Monday next, and I am thinking of being off the Monday following (Ap. 30). I have come to the same conclusion as yourself, that Glion would be better than Grindelwald. I should like very much to see you. Just drop me a line to say when you are likely to turn up.

Poor Arnold's death [1] has been a great shock—rather for his wife than himself—I mean on her account than his. I have always thought sudden death to be the best of all for oneself, but under such circumstances it is terrible for those who are left. Arnold told me years ago that he had heart disease. I do not suppose there is any likelihood of an immediate catastrophe in my own case. I should not go abroad if there were. Imagine the horror of leaving one's wife to fight all the difficulties of sudden Euthanasia in a Swiss hotel! I saw enough of that two years ago at Arolla.—Ever yours,   T. H. HUXLEY.

4 MARLBOROUGH PLACE,
April 25, 1888.

MY DEAR HOOKER—All my beautiful Swiss plans are knocked on the head—at any rate for the present—in favour of horizontality and Digitalis here. The journey up on Monday demonstrated that travelling, at present, was impracticable.

Hames is sanguine I shall get right with rest, and I am quite satisfied with his opinion, but for the sake of my belongings he thinks it right to have Clark's opinion to fortify him.

It is a bore to be converted into a troublesome invalid even for a few weeks, but I comfort myself with my usual reflection on the chances of life, "Lucky it is no worse." Any impatience would have been checked by what I heard about Moseley this morning—that he has sunk into hopeless idiocy. A man in the prime of life!—Ever yours,

T. H. HUXLEY.

---

[1] Matthew Arnold died suddenly of heart disease at Liverpool, where he had gone to meet his daughter on her return from America.

4 MARLBOROUGH PLACE,
May 4, 1888.

MY DEAR HOOKER— Best thanks for your note and queries.

I remember hearing what you say about Darwin's father long ago, I am not sure from what source. But if you look at p. 20 of the *Life and Letters* you will see that D. himself says his father's mind " was not scientific." I have altered the passage so as to use these exact words.

I used "malice" rather in the French sense, which is more innocent than ours, but " irony " would be better if " malice " in any way suggests malignity. " Chaff " is unfortunately beneath the dignity of an R.S. obituary.

I am going to add a short note about Erasmus Darwin's views.

It is a great comfort to me that you like the thing. I am getting nervous over possible senility—63 to-day, and nothing of your evergreen ways about me.

I am decidedly mending, chiefly to all appearance by allowing myself to be stuffed with meat and drink like a Strasburg goose. I am also very much afraid that abolishing tobacco has had something to do with my amendment.

But I am mindful of your maxim—keep a tight hold over your doctor.—Ever yours very faithfully,

T. H. HUXLEY.

P.S. 1.—Can't say I have sacrificed anything to penmanship, and am not at all sure about lucidity !

P.S. 2.—It *is* " Friday "—there is a dot over the i— reopened my letter to crow !

The following letter to Mr. Spencer is in answer to a note of condolence on his illness, in which the following passage occurs :—

I was grieved to hear of so serious an evil as that which [Hirst] named. It is very depressing to find one's friends as well as one's self passing more and more into invalid life.

Well, we always have one consolation, such as it is, that we have made our lives of some service in the world, and that, in fact, we are suffering from doing too much for our fellows. Such thoughts do not go far in the way of mitigation, but they are better than nothing.

4 MARLBOROUGH PLACE,
May 8, 1888.

MY DEAR SPENCER—I have been on the point of writing to you, but put it off for lack of anything cheerful to say.

After I had recovered from my pleurisy, I could not think why my strength did not come back. It turns out that there is some weakness and dilatation of the heart, but luckily no valvular mischief. I am condemned to the life of a prize pig—physical and mental idleness, and corporeal stuffing with meat and drink, and I am certainly improving under the regimen.

I am told I have a fair chance of getting all right again. But I take it as a pretty broad hint to be quiet for the rest of my days. At present I have to be very quiet, and I spend most of my time on my back.

You and I, my dear friend, have had our innings, and carry our bats out while our side is winning. One could not reasonably ask for more. And considering the infinite possibilities of physical and moral suffering which beset us, I, for my part, am well pleased that things are no worse.
Ever yours very faithfully, T. H. HUXLEY.

4 MARLBOROUGH PLACE, N.W.,
June 1, 1888.

MY DEAR KNOWLES—I have been living the life of a prize pig for the last six weeks—no exercise, much meat and drink, and as few manifestations of intelligence as possible, for the purpose of persuading my heart to return to its duty.

I am astonished to find that there is a kick left in me —even when your friend Kropotkin pitches into me without the smallest justification.   *Vide* XIX., June, p. 820.

Just look at XIX., February, p. 168.   I say, "*At the present time*, the produce of the soil does not suffice," etc.

I did not say a word about the capabilities of the soil if, as part and parcel of a political and social revolution on the grandest scale, we all took to spade husbandry.

As a matter of fact, I did try to find out a year or two ago, whether the soil of these islands could, under any circumstances, feed its present population with wheat. I could not get any definite information, but I understood Caird to think that it could.

In my argument, however, the question is of no moment.   There must be some limit to the production of food by a given area, and there is none to population.

What a stimulus vanity is—nothing but the vain dislike of being thought in the wrong would have induced me to trouble myself or bore you with this letter.   Bother Kropotkin !

I think his article very interesting and important nevertheless.

I am getting better, but very slowly.—Ever yours very truly,   T. H. HUXLEY.

In reply, Mr. Knowles begged him to come to lunch and a quiet talk, and further suggested, "as an *entirely unbiassed* person," that he ought to answer Kropotkin's errors in the *Nineteenth Century*, and not only in a private letter behind his back.

The answer is as follows :—

4 Marlborough Place, N.W.,
June 3, 1888.

My dear Knowles—Your invitation is tantalising.
I wish I could accept it. But it is now some six weeks
that my excursions have been limited to a daily drive.
The rest of my time I spend on the flat of my back,
eating, drinking, and doing absolutely nothing besides,
except taking iron and digitalis.

I meant to have gone abroad a month ago, but it
turned out that my heart was out of order, and though I
am getting better, progress is slow, and I do not suppose
I shall get away for some weeks yet.

I have neither brains nor nerves, and the very thought
of controversy puts me in a blue funk !

My doctors prophesy good things, as there is no
valvular disease, only dilatation. But for the present I
must subscribe myself (from an editorial point of view)—
Your worthless and useless and bad-hearted friend,

T. H. Huxley.

The British Association was to meet at Plymouth
this year ; and Mr. W. F. Collier (an uncle of John
Collier, his son-in-law) invited Huxley and any friend
of his to be his guest at Horrabridge.

4 Marlborough Place,
June 13, 1888.

My dear Mr. Collier—It would have been a great
pleasure to me to be your guest once more, but the Fates
won't have it this time.

Dame Nature has given me a broad hint that I have
had my innings, and, for the rest of my time, must be
content to look on at the players.

It is not given to all of us to defy the doctors and go
in for a new lease, as I am glad to hear you are doing. I

declare that your open invitation to any friend of mine is the most touching mark of confidence I ever received. I am going to send it to my great ally Michael Foster, Secretary of the Royal Society.   I do not know whether he has made any other arrangements, and I am not quite sure whether he and his wife are going to Plymouth. But I hope they may be able to accept, for you will certainly like them, and they will certainly like you.   I will ask him to write directly to you to save time.

With very kind remembrances to Mrs. Collier—Ever yours very faithfully,   T. H. HUXLEY.

I forgot to say that I am mending as fast as I can expect to do.

CHAPTER IV

1888

IT was not till June 23 that Huxley was patched up sufficiently by the doctors for him to start for the Engadine. His first stage was to Lugano ; the second by Menaggio and Colico to Chiavenna ; the third to the Maloja. The summer visitors who saw him arrive so feeble that he could scarcely walk a hundred yards on the level, murmured that it was a shame to send out an old man to die there. Their surprise was the greater when, after a couple of months, they saw him walking his ten miles and going up two thousand feet without difficulty. As far as his heart was concerned, the experiment of sending him to the mountains was perfectly justified. With returning strength he threw himself once more into the pursuit of gentians, being especially interested in their distribution and hybridism, and the possibility of natural hybrids explaining the apparent connecting links between species. No doubt, too, he felt some gratification in learning from his friend Mr. (now Sir W.) Thiselton Dyer, that the results he had already obtained in pursuing this hobby had been of real value :—

Your important paper "On Alpine Gentians" (writes the latter) has begun to attract the attention of botanists. It has led Baillon, who is the most acute of the French people, to make some observations of his own.

At the Maloja he stayed twelve weeks, but it was not until nearly two months had elapsed that he could write of any decided improvement, although even then his anticipations for the future were of the gloomiest. The "secret" alluded to in the following letter is the destined award to him of the Copley medal :—

HOTEL KURSAAL, MALOJA,
OBER ENGADINE, *Aug.* 17, 1888.

MY DEAR FOSTER—I know you will be glad to hear that, at last, I can report favourably of my progress. The first six weeks of our stay here the weather was cold, foggy, wet, and windy—in short, everything that it should not be. If the hotel had not been as it is, about the most comfortable in Switzerland, I do not know what I should have done. As it was, I got a very bad attack of "liver," which laid me up for ten days or so. A Brighton doctor—Bluett by name, and well up to his work—kindly looked after me.

With the early days of August the weather changed for the better, and for the last fortnight we have had perfect summer—day after day. I soon picked up my walking power, and one day got up to Lake Longhin, about 2000 feet up. That was by way of an experiment, and I was none the worse for it, but usually my walks are of a more modest description. To-day we are all clouds and rain, and my courage is down to zero, with præcordial discomfort. It seems to me that my heart is quite strong enough to do all that can reasonably be required of it—if all the rest of the machinery is in good order, and the outside conditions are favourable. But the poor old pump cannot contend with grit or want of oil anywhere.

I mean to stay here as long as I can; they say it is often very fine up to the middle of September. Then we shall migrate lower, probably on the Italian side, and get home most likely in October. But I really am very much puzzled to know what to do.

My wife has not been very well lately, and Ethel has contrived to sprain her ankle at lawn-tennis. Collier has had to go to Naples, but we expect him back in a few days.

With our united love to Mrs. Foster and yourself—
Ever yours,   T. H. HUXLEY.

I was very pleased to hear of a secret my wife communicated to me. So long as I was of any use, I did not care much about having the fact recognised, but now that I am used up I like the feather in my cap. "Fuimus." Let us have some news of you.

Sir M. Foster, who was kept in England by the British Association till September 10, wrote that he was going abroad for the rest of September, and proposed to spend some time at Menaggio, whence he hoped to effect a meeting. He winds up with a jest at his recent unusual occupation :—"I have had no end of righteousness accounted to me for helping to entertain Bishops at Cambridge. Hence the postscript in reply : —

HOTEL KURSAAL, MALOJA,
Sept. 2, 1888.

MY DEAR FOSTER—A sharp fall of snow has settled our minds, which have been long wavering about future plans, and we leave this for Menaggio, Hotel Vittoria, on Thursday next, 6th.[1]

All the wiseacres tell us that there are fresher breezes (vento di Lecco) at Menaggio than anywhere else in the Como country, and at anyrate we are going to try whether we can exist there. If it does not answer, we will leave a note for you there to say where we are gone. It would be very jolly to forgather.

I am sorry to leave this most comfortable of hotels, but I do not think that cold would suit either of us. I am marvellously well so long as I am taking sharp exercise, and I do my nine or ten miles without fatigue. It is only when I am quiet that I know that I have a heart.

I do not feel at all sure how matters may be 4000 feet lower, but what I have gained is all to the good in the way of general health. In spite of all the bad weather we have had, I have nothing but praise for this place—the air is splendid, excellent walks for invalids, capital drainage, and the easiest to reach of all places 6000 feet up.

My wife sends her love, and thanks Mrs. Foster for her letter, and looks forward to meeting her.—Ever yours,　　　　　　　　　　　T. H. HUXLEY.

Wash yourself clean of all that episcopal contamination or you may infect me !

But adverse circumstances prevented the meeting.

> HOTEL KURSAAL, MALOJA,
> *Sept.* 24, 1888.

MY DEAR FOSTER—As ill luck would have it, we went over to Pont Resina to day (for the first time), and have only just got back (5.30). I have just telegraphed to you.

---

[1] He did not ultimately leave till the 22nd.

All our plans have been upset by the Föhn wind, which gave us four days' continuous downpour here—upset the roads, and flooded the Chiavenna-Colico Railway. We hear that the latter is not yet repaired.

I was going to write to you at the Vittoria, but thought you could have hardly got there yet. We took rooms there a week ago, and then had to countermand them. If there are any letters kicking about for us, will you ask them to send them on?

By way of an additional complication, my poor wife gave herself an unlucky strain this morning, and even if the railway is mended I do not think she will be fit to travel for two or three days. We are very disappointed. What is to be done?

I am wonderfully better. So long as I am taking active exercise and the weather is dry, I am quite comfortable, and only discover that I have a heart when I am kept quiet by bad weather or get my liver out of order. Here I can walk nine or ten miles up hill and down dale without difficulty or fatigue. What I may be able to do elsewhere is doubtful.—Ever yours,

T. H. HUXLEY.

It would do you and Mrs. Foster a great deal of good to come up here. Not out of your way at all! Oh dear no!

ZÜRICH, *Oct.* 4, 1888.

MY DEAR FOSTER—I should have written to you at Stresa, but I had mislaid your postcard, and it did not turn up till too late.

We made up our minds after all that we would as soon not go down to the Lakes—where the ground would be drying up after the inundations—so we went the other way over the Julier to Tiefenkasten, and from T. to Ragatz, where we stayed a week. Ragatz was hot and steamy at first—cold and steamy afterwards—but earlier in the season, I should think, it would be pleasant.

Last Monday we migrated here, and have had the vilest weather until to-day. All yesterday it rained cats and dogs.

To-day we are off to Neuhausen (Schweitzerhof) to have a look at the Rhine falls. If it is pleasant we may stop there a few days. Then we go to Stuttgart, on our way to Nuremberg, which neither of us have seen. We shall be at the "Bavarian Hotel," and a letter will catch us there, if you have anything to say, I daresay up to the middle of the month. After that Frankfort, and then home.

We do not find long railway journeys very good for either of us, and I am trying to keep within six hours at a stretch.

I am not so vigorous as I was at Maloja, but still infinitely better than when I left England.

I hope the mosquitoes left something of you in Venice. When I was there in October there were none !

My wife joins with me in love to Mrs. Foster and yourself.—Ever yours very faithfully,

T. H. HUXLEY.

Some friendly chaff in Sir M. Foster's reply to the latter contains at least a real indication of the way in which Huxley became the centre of the little society at the Maloja :—

You may reflect that you have done the English tourists a good service this summer. At most *table d'hôtes* in the Lakes I overheard people talking about the joys of Maloja, and giving themselves great airs on account of their intimacy with "Professor Huxley" ! !

But indeed he made several friends here, notably one in an unexpected quarter. This was Father Steffens, Professor of Palæography in Freiburg

University, resident Catholic priest at Maloja in the summer, with whom he had many discussions, and whose real knowledge of the critical questions confronting Christian theology he used to contrast with the frequent ignorance and occasional rudeness of the English representatives of that science who came to the hotel.

A letter to Mr. Spencer from Ragatz shows him on his return journey :—

In fact, so long as I was taking rather sharp exercise in sunshine I felt quite well, and I could walk as well as any time these ten years. It needed damp cold weather to remind me that my pumping apparatus was not to be depended upon under unfavourable conditions. Four thousand feet descent has impressed that fact still more forcibly upon me, and I am quite at sea as to what it will be best to do when we return. Quite certainly, however, we shall not go to Bournemouth. I like the place, but the air is too soft and moist for either of us.

I should be very glad if we could be within reach of you and help to cheer you up, but I cannot say anything definite at present about our winter doings. . . .

My wife sends her kindest regards. She is much better than when we left, which is lucky for me, as I have no mind, and could not make it up if I had any. The only vigour I have is in my legs, and that only when the sun shines.—Ever yours,

T. H. HUXLEY.

A curious incident on this journey deserves recording, as an instance of a futile "warning." On the night of October 6-7, Huxley woke in the night and seemed to hear an inward voice say, "Don't go to Stuttgart and Nuremberg; go straight home." All he did was to make a note of the occurrence and carry out his original plan, whereupon nothing happened.

The following to his youngest daughter, who had gone back earlier from the Maloja, refers to her success in winning the prize for modelling at the Slade School of Art.

SCHWEITZERHOF, NEUHAUSEN,
Oct. 7, 1888.

DEAREST BABS—I will sit to you like "Pater on a monument smiling at grief" for the medallion. As to the photographs, I will try to get them done to order either at Stuttgart or Nuremberg, if we stay at either place long enough. But I am inclined to think they had better be done at home, and then you could adjust the length of the caoutchouc visage to suit your artistic convenience.

We have been crowing and flapping our wings over the medal and trimmings. The only thing I lament is that "your father's influence" was not brought to bear; there is no telling what you might have got if it had been. Thoughtless—very ! !

So sorry we did not come here instead of stopping at Ragatz. The falls are really fine, and the surrounding country a wide tableland, with the great snowy peaks of the Oberland on the horizon. Last evening we had a brilliant sunset, and the mountains were lighted up with the most delicate rosy blush you can imagine.

To-day it rains cats and dogs again. You will have seen in the papers that the Rhine and the Aar and the

Rhone and the Arve are all in flood. There is more water here in the falls than there has been these ten years. However, we have got to go, as the hotel shuts up to-morrow, and there seems a good chance of reaching Stuttgart without water in the carriage.

Long railway journeys do not seem to suit either of us, and we have fixed the maximum at six hours. I expect we shall be home some time in the third week of this month.

Love to Hal and anybody else who may be at home.—Ever your   PATER.

> 4 MARLBOROUGH PLACE,
> *Oct.* 20, 1888.

MY DEAR FOSTER—We got back on Thursday, and had a very good passage, and took it easy by staying the night at Dover. The "Lord Warden" gave us the worst dinner we have had for four months, at double the price of the good dinners. I wonder why we cannot manage these things better in England.

We are both very glad to be at home again, and trust we may be allowed to enjoy our own house for a while. But, oh dear, the air is not Malojal! not even at Hampstead, whither I walked yesterday, and the pump labours accordingly.

I found the first part of the fifth edition of the Text-book among the two or three cwt. of letters and books which had accumulated during four months. Gratulire !

By the way, S. K. has sent me some inquiry about Examinations, which I treat with contempt, as doubtless you have a duplicate.—Ever yours very faithfully,
T. H. HUXLEY.

On October 25 he announces his return to Sir Joseph Hooker, and laments his loss of vigour at the sea-level :—

Hames won't let me stay here in November, and I think we shall go to Brighton.   Unless on the flat of my back, in bed, I shall not have been at home a month all this year.

I have been utterly idle.   There was a lovely case of hybridism, *Gentiana lutea* and *G. punctata,* in a little island in the lake of Sils ; but I fell ill and was confined to bed just after I found it out.   It would be very interesting if somebody would work out Distribution five miles round the Maloja as a centre.   There are the most curious local differences.

You asked me to send you a copy of my obituary of Darwin.   So I put one herewith, though no doubt you have seen it in *Proc. R. S.*

I should like to know what you think of xvii.-xxii. If ever I am able to do anything again I will enlarge on these heads.

In these pages of the Obituary Notice (*Proc. Roy. Soc.* XLIV., No. 269) he endeavours

to separate the substance of the theory from its accidents, and to show that a variety, not only of hostile comments, but of friendly would-be improvements lose their *raison d'être* to the careful student. . . .

It is not essential to Darwin's theory that anything more should be assumed than the facts of heredity, variation, and unlimited multiplication ; and the validity of the deductive reasoning as to the effect of the last (that is, of the struggle for existence which it involves) upon the varieties resulting from the operation of the former. Nor is it essential that one should take up any particular position in regard to the mode of variation, whether, for example, it takes place *per saltum* or gradually ; whether it is definite in character or indefinite.   Still less are those who accept the theory bound to any particular views as to the causes of heredity or of variation.

The remaining letters of the year trace the gradual bettering of health, from the "no improvement" of October to the almost complete disappearance of bad symptoms in December. He had renounced Brighton, which he detested, in favour of Eastbourne, where the keen air of the downs and the daily walk over Beachy Head acted as a tolerable substitute for the Alps. Though he would not miss the anniversary meeting of the Royal Society, when he was to receive the Copley medal, one more link binding him to his old friend Hooker, he did not venture to stay for the dinner in the evening.

This autumn also he resigned his place on the board of Governors of Eton College. "I think it must be a year and a half," he writes, "since I attended a meeting, and I am not likely to do better in the future."

4 MARLBOROUGH PLACE,
Oct. 28, 1888.

MY DEAR HOOKER—Best thanks for your suggestion about the cottage, viz. "that before you decide on Brighton Mrs. Huxley should come down and look at the cottage below my house" at Sunningdale, but I do not see my way to adopting it. A house, however small, involves servants and ties one to one place. The conditions that suit me do not seem to be found anywhere but in the high Alps, and I can't afford to keep a second house in the country and pass the summer in Switzerland as well.

We are going to Brighton (not because we love it, quite t'other) on account of the fine weather that is to be had there in November and December. We shall be back for some weeks about Christmas, and then get away somewhere else—Malvern possibly—out of the east winds of February and March.

I do not like this nomadic life at all, but it appears to be Hobson's choice between that and none.

I am sorry to hear you are troubled by your ears. I am so deaf that I begin to fight shy of society. It irritates me not to hear; it irritates me still more to be spoken to as if I were deaf, and the absurdity of being irritated on the last ground irritates me still more.

I wish you would start that business of giving a competent young botanist with good legs £100 to go and study distribution in the Engadine—from the Maloja as centre—in a circle of a radius of eight or ten miles. The distribution of the four principal conifers, Arolla pine, larch, mountain pine and spruce, is most curious, the why and wherefore nowise apparent.

I am very sorry I cannot be at x on Thursday, but they won't let me be out at night at present.—Ever yours,

T. H. Huxley.

4 Marlborough Place,
Oct. 28, 1888.

My dear Foster—No fear of my trying to stop in London. Hames won't have it. He came and over-hauled me the other day. As I expected, the original mischief is just as it was. One does not get rid either of dilatation or its results at my time of life. The only thing is to keep the pipes clear by good conditions of existence.

After endless discussion we have settled on Brighton for November and December. It is a hateful place to my mind, but there is more chance of sunshine there (at this time) than anywhere else. We shall come up for a week or two on this side of Christmas, and then get away somewhere else out of the way of the east winds of February and March.

I do not think that the Hazlemere country would do
for us, nor indeed any country place so long as we cannot
regularly set up house.

Heaven knows I don't want to bother about anything
at present. But I should like to convince —— that he
does not yet understand the elements of [his subject].
What a copious inkspilling cuttlefish of a writer he is !—
Ever yours,   T. H. HUXLEY.

4 MARLBOROUGH PLACE, LONDON, N.W.,
Nov. 2, 1888.

MY DEAR SKELTON—Best thanks for the second volume
of *Maitland of Lethington.* I have been in the Engadine
for the last four months, trying to repair the crazy old
" house I live in," and meeting with more success than I
hoped for when I left home.

Your volume turned up amidst a mountain of accumulated books, papers, and letters, and I can only hope
it has not been too long without acknowledgment.

I have been much interested in your argument about
the " Casket letters." The comparison of Crawford's
deposition with the Queen's letter leaves no sort of doubt
that the writer of one had the other before him ; and
under the circumstances I do not see how it can be
doubted that the Queen's letter is forged.

But though thus wholly agreeing with you in substance, I cannot help thinking that your language on
p. 341 may be seriously pecked at.

My experience of reporters leads me to think that
there would be no discrepancy at all comparable to that
between the two accounts, and I speak from the woeful
memories of the many Royal Commissions I have wearied
over. The accuracy of a good modern reporter is really
wonderful.

And I do not think that " the two documents
were drawn by the same hand." I should say that the

writer of the letter had Crawford's deposition before him, and made what he considered improvements here and there.

You will say this letter is like Falstaff's reckoning, with but a pennyworth of thanks to this monstrous quantity of pecking.

But the gratitude is solid and the criticism mere two dimension stuff.   It is a charming book.

With   kind   remembrances to   Mrs.   Skelton — Ever yours very faithfully,   T. H. HUXLEY.

                    10 SOUTHCLIFF TERRACE, EASTBOURNE,
                              Nov. 9, 1888.

MY DEAR FOSTER—We came here on Tuesday, on which day, by ill luck, the east wind also started, and has been blowing half a gale ever since.   We are in the last house but one to the west, and as high up as we dare go —looking out on the sea.   The first day we had to hold on to our chairs to prevent being blown away in the sitting-room, but we have hired a screen and can now croon over the fire without danger.

A priori, the conditions cannot be said to have been promising for two people, one of whom is liable to bronchitis and rheumatism and the other to pleurisy, but, as I am so fond of rubbing into Herbert Spencer, *a priori* reasonings are mostly bosh, and we are thriving.

With three coats on I find the air on Beachy Head eminently refreshing, and there is so much light in the southern quarter just now, that we confidently hope to see the sun once more in the course of a few days.

As I told you in my official letter, I am going up for the 30th.   But I am in a quandary about the dinner, partly by reason of the inevitable speech, and partly the long sitting.   I should very much like to attend, and I think I could go through with it.   On the other hand, my wife declares it would be very imprudent, and I am not quite sure she is wrong. I wish you would tell me exactly what you think about the matter.

The way I pick up directly I get into good air makes me suspect myself of malingering, and yet I certainly had grown very seedy in London before we left.—Ever yours, T. H. HUXLEY.

10 SOUTHCLIFF TERRACE, EASTBOURNE,
Nov. 13, 1888.

MY DEAR FOSTER—We are very sorry to hear about Michael Junior.[1] *Experto crede;* of all anxieties the hardest to bear is that about one's children. But considering the way you got off yourself and have become the hearty and bucolic person you are, I think you ought to be cheery. Everybody speaks well of the youngster, and he is bound to behave himself well and get strong as swiftly as possible.

Though very loth, I give up the dinner. But unless I am on my back I shall turn up at the meeting. I think that is a compromise very creditable to my prudence.

Though it is blowing a gale of wind from S.W. to-day there is real sunshine, and it is fairly warm. I am very glad we came here instead of that beastly Brighton.—Ever yours very faithfully, T. H. HUXLEY.

10 SOUTHCLIFF TERRACE, EASTBOURNE,
Nov. 15, 1888.

MY DEAR EVANS—I am very sorry to have missed you. I told my doctor that while the weather was bad it was of no use to go away, and when it was fine I might just as well stop at home; but he did not see the force of my reasoning, and packed us off here.

---

[1] Sir M. Foster s son was threatened with lung trouble, and was ordered to live abroad. He proposed to carry his medical experience to the Maloja and practise there during the summer. Huxley offered to give him some introductions.

The award of the Copley is a kindness I feel very
much. . . .

The Congress [1] seems to have gone off excellently.   I
consider that my own performance of the part of dummy
was distinguished.

So the Lawes business is fairly settled at last !   " Lawes
Deo," as the Claimant might have said.   But the pun
will be stale, as you doubtless have already made all
possible epigrams and punnigrams on the topic.

My wife joins with me in kindest regards to Mrs.
Evans and yourself.   If Mrs. Evans had only come up to
the Maloja, she would have had real winter and no cold.
—Ever yours very faithfully,   T. H. HUXLEY.

10 SOUTHCLIFF TERRACE, EASTBOURNE,
Nov. 15, 1888.

MY DEAR HOOKER—You would have it that the R.S.
broke the law in giving you the Copley, and they certainly
violated custom in giving it to me the year following.
Who ever heard of two biologers getting it one after
another ?   It is very pleasant to have our niches in the
Pantheon close together.   It is getting on for forty years
since we were first " acquent," and considering with what
a very considerable dose of tenacity, vivacity, and that
glorious firmness (which the beasts who don't like us
call obstinacy) we are both endowed, the fact that we
have never had the shadow of a shade of a quarrel is
more to our credit than being ex-Presidents and Copley
medallists.

But we have had a masonic bond in both being well
salted in early life.   I have always felt I owed a great
deal to my acquaintance with the realities of things gained
[in] the old Rattlesnake.

I am getting on pretty well here, though the weather

---

[1] The International Geological Congress, at which he was to
have presided.

has been mostly bad.   All being well I shall attend the
meeting of the Society on the 30th, but not the dinner.
I am very sorry to miss the latter, but I dare not face
the fatigue and the chances of a third dose of pleurisy.

My wife sends kindest regards and thanks for your
congratulations.—Ever yours very faithfully,

T. H. HUXLEY.

10 SOUTHCLIFF TERRACE, EASTBOURNE,
Nov. 17, 1888.

MY DEAR FLOWER— . . . Many thanks for taking my
troublesomeness in good part.   My friend will be greatly
consoled to know that you have the poor man " in your
eye."   Schoolmaster, naturalist, and coal merchant used to
be the three refuges for the incompetent.   Schoolmaster
is rapidly being eliminated, so I suppose the pressure on
Natural History and coals will increase.

I am glad you have got the Civil Service Commissioners
to listen to common sense.   I had an awful battle with
them (through the Department) over Newton, who is now
in your paleontological department.   If I recollect rightly,
they examined him *inter alia* on the working of the Poor
Laws !

The Royal Society has dealt very kindly with me.
They patted me on the back when I started thirty-seven
years ago, and it was a great encouragement.   They give
me their best, now that my race is run, and it is a great
consolation.   At the far end of life all one's work looks so
uncommonly small, that the good opinion of one's contemporaries acquires a new value.

We have a summer's day, and I am writing before an
open window !   Yesterday it blew great guns.—Ever
yours very faithfully,   T. H. HUXLEY.

The following letter to Lady Welby, the point of
which is that to be " morally convinced " is not the

same thing as to offer scientific proof, refers to an article in the *Church Quarterly* for October called "Truthfulness in Science and Religion," evoked by Huxley's *Nineteenth Century* article on "Science and the Bishops."

Nov. 27, 1888.

DEAR LADY WELBY—Many thanks for the article in the *Church Quarterly*, which I return herewith. I am not disposed to bestow any particular attention upon it; as the writer, though evidently a fair-minded man, appears to me to be entangled in a hopeless intellectual muddle, and one which has no novelty. Christian beliefs profess to be based upon historical facts. If there was no such person as Jesus of Nazareth, and if His biography given in the Gospels is a fiction, Christianity vanishes.

Now the inquiry into the truth or falsehood of a matter of history is just as much a question of pure science as the inquiry into the truth or falsehood of a matter of geology, and the value of evidence in the two cases must be tested in the same way. If any one tells me that the evidence of the existence of man in the miocene epoch is as good as that upon which I frequently act every day of my life, I reply that this is quite true, but that it is no sort of reason for believing in the existence of miocene man.

Surely no one but a born fool can fail to be aware that we constantly, and in very grave conjunctions, are obliged to act upon extremely bad evidence, and that very often we suffer all sorts of penalties in consequence. And surely one must be something worse than a born fool to pretend that such decision under the pressure of the enigmas of life ought to have the smallest influence in those judgments which are made with due and sufficient deliberation. You will see that these considerations go to the root of the whole matter. I regret that I cannot discuss the question more at length and deal with sundry topics put forward in your letter. At present writing is a burden to me.

A letter to Professor Ray Lankester mixes grave and gay in a little homily, edged by personal experience, on the virtues and vices of combativeness.

10 SOUTHCLIFF TERRACE, EASTBOURNE,
Dec. 6, 1888.

I think it would be a very good thing both for you and for Oxford if you went there. Oxford science certainly wants stirring up, and notwithstanding your increase in years and wisdom, I think you would bear just a little more stirring down, so that the conditions for a transfer of energy are excellent !

Seriously, I wish you would let an old man, who has had his share of fighting, remind you that battles, like hypotheses, are not to be multiplied beyond necessity. Science might say to you as the Staffordshire collier's wife said to her husband at the fair, "Get thee foighten done and come whoam." You have a fair expectation of ripe vigour for twenty years; just think what may be done with that capital.

No use to *tu quoque* me. Under the circumstances of the time, warfare has been my business and duty.—Ever yours very faithfully,   T. H. HUXLEY.

Two more letters of the year refer to the South Kensington examinations, for which Huxley was still nominally responsible. As before, we see him reluctant to sign the report upon papers which he had not himself examined; yet at the same time doing all that lay in his power to assist by criticising the questions and thinking out the scheme of teaching on which the examination was to be based. He replies to some proposed changes in a letter to Sir M. Foster of December 12 :—

I am very sorry I cannot agree with your clients about the examination. They should recollect the late Master of Trinity's aphorism that even the youngest of us is not infallible.

I know exactly upon what principles I am going, and so far as I am at present informed that advantage is peculiar to my side. Two points I am quite clear about— one is the exclusion of *Amphioxus*, and the other the retention of so much of the Bird as will necessitate a knowledge of Sauropsidan skeletal characters and the elements of skeletal homologies in skull and limbs.

I have taken a good deal of pains over drawing up a new syllabus—including dogfish—and making room for it by excluding *Amphioxus* and all of bird except skeleton. I have added Lamprey (cranial and spinal skeleton, *not* face cartilages), so that the intelligent student may know what a notochord means before he goes to embryology. I have excluded *Distoma* and kept *Helix*.

The Committee must now settle the matter. I have done with it.

On December 27 he writes —

I have been thinking over the Examinership business without coming to any very satisfactory result. The present state of things is not satisfactory so far as I am concerned. I do not like to appear to be doing what I am not doing.

—— would of course be the successor indicated, if he had not so carefully cut his own throat as an Examiner. . . . He would be bringing an action against the Lord President before he had been three years in office ! . . . As I told Forster, when he was Vice-President, the whole value of the Exr. system depends on the way the examiners do their work. I have the gravest doubt about —— steadily plodding through the disgustful weariness of it as you and I have done, or observing any regulation that did not suit his fancy.

With this may be compared the letter of May 19, 1889, to Sir J. Donnelly, when he finally resolved to give up the "sleeping partnership" in the examination.

His last letter of the year was written to Sir J. Hooker, when transferring to him the "archives" of the *x* Club, as the new Treasurer.

4 MARLBOROUGH PLACE,
Dec. 29, 1888.

MY DEAR HOOKER—All good wishes to you and yours, and many of 'em.

Thanks for the cheque. You are very confiding to send it without looking at the account. But I have packed up the "Archives," which poor dear Busk handed over to me, and will leave them at the Athenæum for you. Among them you will find the account book. There are two or three cases, when I was absent, in which the names are not down. I have no doubt Frankland gave them to me by letter, but the book was at home and they never got set down. *Peccavi!*

I have been picking up in the most astonishing way during the last fortnight or three weeks at Eastbourne. My doctor, Hames, carefully examined my heart yesterday, and told me that though some slight indications were left, he should have thought nothing of them if he had not followed the whole history of the case. With fresh air and exercise and careful avoidance of cold and night air I am to be all right again in a few months.

I am not fond of coddling; but as Paddy gave his pig the best corner in his cabin—because "shure, he paid the rint"—I feel bound to take care of myself as a household animal of value, to say nothing of any other grounds. So, much as I should like to be with you all on the 3rd, I must defer to the taboo.

The wife got a nasty bronchitic cold as soon as she came up. She is much better now. But I shall be glad to get her down to Eastbourne again.

Except that, we are all very flourishing, as I hope you are.—Ever yours very faithfully,   T. H. HUXLEY.

CHAPTER V

1889

THE events to be chronicled in this year are, as might be expected, either domestic or literary. The letters are full of allusions to his long controversy in defence of Agnosticism, mainly with Dr. Wace, who had declared the use of the name to be a "mere evasion" on the part of those who ought to be dubbed infidels;[1] to the building of the new house at Eastbourne, and

---

[1] Apropos of this controversy, a letter may be cited which appeared in the *Agnostic Annual* for 1884, in answer to certain inquiries from the editor as to the right definition of Agnosticism :—

Some twenty years ago, or thereabouts, I invented the word "Agnostic" to denote people who, like myself, confess themselves to be hopelessly ignorant concerning a variety of matters, about which metaphysicians and theologians, both orthodox and heterodox, dogmatise with the utmost confidence, and it has been a source of some amusement to me to watch the gradual acceptance of the term and its correlate, "Agnosticism" (I think the *Spectator* first adopted and popularised both), until now Agnostics are assuming the position of a recognised sect, and Agnosticism is honoured by especial obloquy on the part of the orthodox. Thus it will be seen that I have a sort of patent right in "Agnostic" (it is my trade mark), and I am entitled to say that I can state authentically what was originally meant by Agnosticism. What other people may understand by it, by this time, I do not know. If a General

to the marriage in quick succession of his two
youngest daughters, whereby, indeed, the giving up
of the house in London and definite departure from
London was made possible.

All the early part of the year, till he found it
necessary to go to Switzerland again, he stayed
unwillingly in Eastbourne, from time to time running
up to town, or having son or daughter to stay with
him for a week, his wife being too busy to leave town,
with the double preparations for the weddings on
hand, so that he writes to her : "I feel worse than the
'cowardly agnostic' I am said to be—for leaving you
to face your botherations alone." One can picture

Council of the Church Agnostic were held, very likely I should be
condemned as a heretic. But I speak only for myself in answering
these questions.

1. Agnosticism is of the essence of science, whether ancient or
modern. It simply means that a man shall not say he knows or
believes that which he has no scientific grounds for professing to
know or believe.

2. Consequently Agnosticism puts aside not only the greater
part of popular theology, but also the greater part of popular anti-
theology. On the whole, the "bosh" of heterodoxy is more
offensive to me than that of orthodoxy, because heterodoxy pro-
fesses to be guided by reason and science, and orthodoxy does not.

3. I have no doubt that scientific criticism will prove destructive
to the forms of supernaturalism which enter into the constitution
of existing religions. On trial of any so-called miracle the verdict
of science is "Not proven." But true Agnosticism will not forget
that existence, motion, and law-abiding operation in nature are
more stupendous miracles than any recounted by the mythologies,
and that there may be things, not only in the heavens and earth,
but beyond the intelligible universe, which "are not dreamt of in
our philosophy." The theological "gnosis" would have us believe
that the world is a conjurer's house ; the anti-theological "gnosis"
talks as if it were a "dirt-pie," made by the two blind children,
Law and Force. Agnosticism simply says that we know nothing
of what may be behind phenomena.

him still firm of tread, with grizzled head a little
stooped from his square shoulders, pacing the sea
wall with long strides, or renewing somewhat of his
strength as it again began to fail, in the keener air of
the downs, warmly defended against chill by a big
cap—for he had been suffering from his ears—and a
long rough coat. He writes (February 22): "I have
bought a cap with flaps to protect my ears. I look
more "doggy' than ever." And on March 3 :—

We have had a lovely day, quite an Italian sky and
sea, with a good deal of Florentine east wind. I walked
up to the Signal House, and was greatly amused by a
young sheep-dog whose master could hardly get him
away from circling round me and staring at me with a
short dissatisfied bark every now and then. It is the
undressed wool of my coat bothers all the dogs. They
can't understand why a creature which smells so like a
sheep should walk on its hind legs. I wish I could have
relieved that dog's mind, but I did not see my way to
an explanation.

From this time on, the effects of several years'
comparative rest became more perceptible. His
slowly returning vigour was no longer sapped by the
unceasing strain of multifarious occupations. And if
his recurrent ill-health sometimes seems too strongly
insisted on, it must be remembered that he had
always worked at the extreme limit of his powers—
the limit, as he used regretfully to say, imposed on
his brain by his other organs—and that after his
first breakdown he was never very far from a second.
When this finally came in 1884, his forces were so

far spent that he never expected to recover as he did.

In the marriage this year of his youngest daughter, Huxley was doomed to experience the momentary little twinge which will sometimes come to the supporter of an unpopular principle when he first puts it into practice among his own belongings.

ATHENÆUM CLUB,
Jan. 14, 1889.

MY DEAR HOOKER—I have just left the *x* "Archives" here for you. I left them on my table by mischance when I came here on the *x* day.

I have a piece of family news for you. My youngest daughter Ethel is going to marry John Collier.

I have always been a great advocate for the triumph of common sense and justice in the "Deceased Wife's Sister" business—and only now discover, that I had a sneaking hope that all of my own daughters would escape that experiment!

They are quite suited to one another and I would not wish a better match for her. And whatever annoyances and social pin-pricks may come in Ethel's way, I know nobody less likely to care about them.

We shall have to go to Norway, I believe, to get the business done.

In the meantime, my wife (who has been laid up with bronchitic cold ever since we came home) and I have had as much London as we can stand, and are off to-morrow to Eastbourne again, but to more sheltered quarters.

I hope Lady Hooker and you are thriving. Don't conceal the news from her, as my wife is always accusing me of doing.—Ever yours,

T. H. HUXLEY.

To Mr. W. F. Collier

4 Marlborough Place,
Jan. 24, 1889.

Many thanks for your kind letter. I have as strong
an affection for Jack as if he were my own son, and I
have felt very keenly the ruin we involuntarily brought
upon him — by our poor darling's terrible illness and
death. So that if I had not already done my best to aid
and abet other people in disregarding the disabilities imposed by the present monstrous state of the law, I should
have felt bound to go as far as I could towards mending
his life. Ethel is just suited to him. . . Of course I
could have wished that she should be spared the petty
annoyances which she must occasionally expect. But I
know of no one less likely to care for them.

Your Shakespere parable[1] is charming—but I am afraid
it must be put among the endless things that are read *in*
to the "divine Williams" as the Frenchman called him.

There was no knowledge of the sexes of plants in
Shakespere's time, barring some vague suggestion about
figs and dates. Even in the 18th century, after Linnæus,
the observations of Sprengel, who was a man of genius,
and first properly explained the action of insects, were
set aside and forgotten.

I take it that Shakespere is really alluding to the
"enforced chastity" of Dian (the moon). The poets
ignore that little Endymion business when they like !

I have recovered in such an extraordinary fashion

---

[1] The second part of the letter replies to the question whether
Shakespeare had any notion of the existence of the sexes in plants
and the part played in their fertilisation by insects, which, of
course, would be prevented from visiting them by rainy weather,
when he wrote in the *Midsummer Night's Dream*—

> The moon, methinks, looks with a watery eye,
> And when she weeps, weeps every little flower
> Lamenting some enforced chastity.

that I can plume myself on being an "interesting case," though I am not going to compete with you in that line. And if you look at the February *Nineteenth* I hope you will think that my brains are none the worse. But perhaps that conceited speech is evidence that they are.

We came to town to make the acquaintance of Nettie's *fiancé*, and I am happy to say the family takes to him. When it does not take to anybody, it is the worse for that anybody.

So, before long, my house will be empty, and as my wife and I cannot live in London, I think we shall pitch our tent in Eastbourne. Good Jack offers to give us a *pied à terre* when we come to town. To-day we are off to Eastbourne again. Carry off Harry, who is done up from too zealous Hospital work. However, it is nothing serious.

The following is in reply to a request that he would write a letter, as he describes it elsewhere, "about the wife's sister business—for the edification of the peers."

3 JEVINGTON GARDENS, EASTBOURNE,
March 12, 1889.

MY DEAR DONNELLY—1 feel "downright mean," as the Yankees say, that I have not done for the sake of right and justice what I am moved to do now that I have a personal interest in the matter of the directest kind ; and I rather expect that will be thrown in my teeth if my name is at the bottom of anything I write.

On the other hand, I loathe anonymity. However, we can take time to consider that point.

Anyhow I will set to work on the concoction of a letter, if you will supply me with the materials which will enable me to be thoroughly posted up in the facts.

I have just received your second letter. Pity you could not stay over yesterday—it was very fine.—Ever yours very faithfully,   T. H. HUXLEY.

The letter in question is as follows :—

April 30, 1889.

DEAR LORD HARTINGTON— I am assured by those who
know more about the political world than I do, that if
Lord Salisbury would hold his hand and let his party do
as they like about the D.W.S. Bill which is to come on
next week, it would pass. Considering the irritation
against the bishops and a certain portion of the lay peers
among a number of people who have the means of making
themselves heard and felt, which is kept up and aggra-
vated, as time goes on, by the action of the Upper House
in repeatedly snubbing the Lower, about this question, I
should have thought it (from a Conservative point of
view) good policy to heal the sore.

The talk of Class *v.* Mass is generally mere clap-trap ;
but, in this case, there is really no doubt that a fraction
of the Classes stands in the way of the fulfilment of a
very reasonable demand on the part of the Masses.

A clear-headed man like Lord Salisbury would surely
see this if it were properly pressed on his attention.

I do not presume to say whether it is practicable or
convenient for the Leader of the Liberal Unionist party
to take any steps in this direction ; and I should hardly
have ventured to ask you to take this suggestion into
consideration if the interest I have always taken in the
D.W S. Bill had not recently been quickened by the
marriage of one of my daughters as a Deceased Wife's
Sister.—I am, etc.

Meantime the effect of Eastbourne, which Sir
John Donnelly had induced him to try, was indeed
wonderful. He found in it the place he had so long
been looking for. References to his health read very
differently from those of previous years. He walked
up Beachy Head regularly without suffering from

any heart symptoms. And though Beachy Head
was not the same thing as the Alps, it made a very
efficient substitute for a while, and it was not till
April that the need of change began to make itself
felt. And so he made up his mind to listen no more
to the eager friends who wished him to pitch his tent
near them at either end of Surrey, but to settle down
at Eastbourne, and, by preference, to build a house
of the size and on the spot that suited himself, rather
than to take any existing house lower down in the
town. He must have been a trifle irritated by un-
solicited advice when he wrote the following :—

> It is very odd that people won't give one credit for
> common sense. We have tried one winter here, and if
> we tried another we should be just as much dependent
> upon the experience of longer residents as ever we were.
> However, as I told X. I was going to settle matters to-
> morrow, there won't be any opportunity for discussing
> that topic when he comes. If we had taken W.'s house,
> somebody would have immediately told us that we had
> chosen the dampest site in winter and the stuffiest in
> summer, and where, moreover, the sewage has to be
> pumped up into the main drain.

He finally decided upon a site on the high ground
near Beachy Head, a little way back from the sea
front, at the corner of the Staveley and Buxton
Roads, with a guarantee from the Duke of Devon-
shire's agent that no house should be built at the
contiguous end of the adjoining plot of land in the
Buxton Road, a plot which he himself afterwards
bought. The principal rooms were planned for the

back of the house, looking S.W. over open gardens to the long line of downs which culminate in Beachy Head, but with due provision against southerly gales and excess of sunshine.

On May 29 the builder's contract was accepted, and for the rest of the year the progress of the house, which was designed by his son-in-law, F. W. Waller, afforded a constant interest.

Meantime, with the improvement in his general health, the old appetite for work returned with increased and unwonted zest. For the first time in his life he declares that he enjoyed the process of writing. As he wrote somewhat later to his newly married daughter from Eastbourne, where he had gone again very weary the day after her wedding: "Luckily the bishops and clergy won't let me alone, so I have been able to keep myself pretty well amused in replying." The work which came to him so easily and pleasurably was the defence of his attitude of agnosticism against the onslaught made upon it at the previous Church Congress by Dr. Wace, the Principal of King's College, London, and followed up by articles in the *Nineteenth Century* from the pen of Mr. Frederic Harrison and Mr. Laing, the effect of which upon him he describes to Mr. Knowles on December 30, 1888 :—

I have been stirred up to the boiling pitch by Wace, Laing, and Harrison *in re* Agnosticism, and I really can't keep the lid down any longer. Are you minded to admit a goring article into the February *Nineteenth?*

As for his health, he adds :—

I have amended wonderfully in the course of the last six weeks, and my doctor tells me I am going to be completely patched up—seams caulked and made seaworthy, so the old hulk may make another cruise.

We shall see. At any rate I have been able and willing to write lately, and that is more than I can say for myself for the first three-quarters of the year.

. . . I was so pleased to see you were in trouble about your house. Good for you to have a taste of it for yourself.

To this controversy he contributed four articles; three directly in defence of Agnosticism, the fourth on the value of the underlying question of testimony to the miraculous.

The first article, " Agnosticism," appeared in the February number of the *Nineteenth Century.* No sooner was this finished than he began a fresh piece of work, "which," he writes, "is all about miracles, and will be rather amusing." This, on the " Value of Testimony to the Miraculous," appeared in the following number of the *Nineteenth Century.* It did not form part of the controversy on hand, though it bore indirectly upon the first principles of agnosticism. The question at issue, he urges, is not the possibility of miracles, but the evidence to their occurrence, and if from preconceptions or ignorance the evidence be worthless the historical reality of the facts attested vanishes. The cardinal point, then, "is completely, as the author of *Robert Elsmere* says, the value of testimony."

The March number also contained replies from
Dr. Wace and Bishop Magee on the main question,
and an article by Mrs. Humphry Ward on a kindred
subject to his own, "The New Reformation." Of
these he writes on February 27 :—

The Bishop and Wace are hammering away in the
Nineteenth. Mrs. Ward's article very good, and practically
an answer to Wace. Won't I stir them up by and by.

And a few days later :—

Mrs. Ward's service consists in her very clear and clever
exposition of critical results and methods.

3 JEVINGTON GARDENS, EASTBOURNE,
Feb. 29, 1889.

MY DEAR KNOWLES —I have just been delighted with
Mrs. Ward's article. She has swept away the greater
part of Wace's sophistries as a dexterous and strong-
wristed housemaid sweeps away cobwebs with her broom,
and saved a lot of time.

What in the world does the Bishop mean by saying
that I have called Christianity "sorry stuff" (p. 370)?
To my knowledge I never so much as thought anything
of the kind, let alone saying it.

I shall challenge him very sharply about this, and if,
as I believe, he has no justification for his statement, my
opinion of him will be very considerably lowered.

Wace has given me a lovely opening by his profession
of belief in the devils going into the swine. I rather
hoped I should get this out of him.

I find people are watching the game with great interest,
and if it should be possible for me to give a little shove
to the "New Reformation," I shall think the fag end of
my life well spent.

After all, the reproach made to the English people
that " they care for nothing but religion and politics" is

rather to their credit. In the long run these are the two things that ought to interest a man more than any others.

I have been much bothered with ear-ache lately, but if all goes well I will send you a screed by the middle of March.

Snowing hard! They have had more snow within the last month than they have known for ten years here. —Ever yours faithfully,   T. H. HUXLEY.

He set to work immediately, and within ten days despatched his second contribution, "Agnosticism, a Rejoinder," which appeared in the April number of the *Nineteenth Century*.

On March 3 he writes :—

I am possessed by a writing demon, and have pretty well finished in the rough another article for Knowles, whose mouth is wide open for it.

And on the 9th :—

I sent off another article to Knowles last night—a regular facer for the clericals. You can't think how I enjoy writing now for the first time in my life.

He writes at greater length to Mr. Knowles :—

3 JEVINGTON GARDENS, EASTBOURNE,
March 10, 1889.

MY DEAR KNOWLES—There's a Divinity that shapes the ends (of envelopes!) rough-hew them how we will. This time I went and bought the strongest to be had, and sealed him up with wax in the shop. I put no note inside, meaning to write to you afterwards, and then I forgot to do so.

I can't understand Peterborough nohow. However, so far as the weakness of the flesh would permit me to abstain from smiting him and his brother Amalekite, I have tried to turn the tide of battle to matters of more importance.

The pith of my article is the proposition that Christ was not a Christian. I have not ventured to state my thesis exactly in that form—fearing the Editor—but, in a mild and proper way, I flatter myself I have demonstrated it. Really, when I come to think of the claims made by orthodox Christianity on the one hand, and of the total absence of foundation for them on the other, I find it hard to abstain from using a phrase which shocked me very much when Strauss first applied it to the Resurrection, " Welthistorischer Humbug ! "

I don't think I have ever seen the portrait you speak of. I remember the artist—a clever fellow, whose name, of course, I forget—but I do not think I saw his finished work. Some of these days I will ask to see it.

I was pretty well finished after the wedding, and bolted here the next day. I am sorry to say I could not get my wife to come with me. If she does not knock up I shall be pleasantly surprised. The young couple are flourishing in Paris. I like what I have seen of him very much.

What is the " Cloister scheme " ? [1] Recollect how far away I am from the world, the flesh and the d——.

Are you and Mrs. Knowles going to imitate the example of Eginhard and Emma ? What good pictures you will have in your monastery church !—Ever yours very faithfully,   T. H. HUXLEY.

And again, a few days later :—

3 JEVINGTON GARDENS, EASTBOURNE,
March 15, 1889.

MY DEAR KNOWLES—I am sending my proof back to Spottiswoode's. I did not think the MS. would make so

---

[1] It referred to a plan for using the cloisters of Westminster Abbey to receive the monuments of distinguished men, so as to avoid the necessity of enlarging the Abbey itself.

much, and I am afraid it has lengthened in the process of correction.

You have a reader in your printer's office who provides me with jokes. Last time he corrected, where my MS. spoke of the pigs as unwilling "porters" of the devils, into "porkers." And this time, when I, writing about the Lord's Prayer, say "current formula," he has it "canting formula." If only Peterborough had got hold of that! And I am capable of overlooking anything in a proof.

You see we have got to big questions now, and if these are once fairly before the general mind all the King's horses and all the King's men won't put the orthodox Humpty Dumpty where he was before.—Ever yours very faithfully,   T. H. HUXLEY.

After the article came out he wrote again to Mr. Knowles :—

4 MARLBOROUGH PLACE, N.W.,
April 14, 1889.

MY DEAR KNOWLES—I am going to try and stop here, desolate as the house is now all the chicks have flown, for the next fortnight. Your talk of the inclemency of Torquay is delightfully consoling. London has been vile.

I am glad you are going to let Wace have another "go." My object, as you know, in the whole business has been to rouse people to think. . . .

Considering that I got named in the House of Commons last night as an example of a temperate and well-behaved blasphemer,[1] I think I am attaining my object.

Of course I go for a last word, and I am inclined to

---

[1] In the debate upon the Religious Prosecutions Abolition Bill, Mr. Addison said "the last article by Professor Huxley in the *Nineteenth Century* showed that opinion was free when it was honestly expressed."—*Times*, April 14.

think that whatever Wace may say, it may be best to get out of the region of controversy as far as possible and hammer in two big nails—(1) that the Demonology of Christianity shows that its founders knew no more about the spiritual world than anybody else, and (2) that Newman's doctrine of "Development" is true to an extent of which the Cardinal did not dream.

I have been reading some of his works lately, and I understand now why Kingsley accused him of growing dishonesty.

After an hour or two of him I began to lose sight of the distinction between truth and falsehood.—Ever yours,

T. H. HUXLEY.

If you are at home any day next week I will look in for a chat.

The controversy was completed by a third article, "Agnosticism and Christianity," in the June number of the *Nineteenth Century.* There was a humorous aspect of this article which tickled his fancy immensely, for he drove home his previous arguments by means of an authority whom his adversaries could not neglect, though he was the last man they could have expected to see brought up against them in this connection—Cardinal Newman. There is no better evidence for ancient than for modern miracles, he says in effect; let us therefore accept the teachings of the Church which maintains a continuous tradition on the subject. But there is a very different conclusion to be drawn from the same premises; all may be regarded as equally doubtful, and so he writes on May 30 to Sir J. Hooker :—

By the way, I want you to enjoy my wind-up with Wace in this month's *Nineteenth* in the reading as much as I have in the writing. It's as full of malice [1] as an egg is full of meat, and my satisfaction in making Newman my accomplice has been unutterable. That man is the slipperiest sophist I have ever met with. Kingsley was entirely right about him.

Now for peace and quietness till after the next Church Congress !

Three other letters to Mr. Knowles refer to this article.

> 4 MARLBOROUGH PLACE, N.W.,
> *May* 4, 1889.

MY DEAR KNOWLES—I am at the end of my London tether, and we go to Eastbourne (3 Jevington Gardens again) on Monday.

I have been working hard to finish my paper, and shall send it to you before I go.

I am astonished at its meekness. Being reviled, I revile not ; not an exception, I believe, can be taken to the wording of one of the venomous paragraphs in which the paper abounds. And I perceive the truth of a profound reflection I have often made, that reviling is often morally superior to not reviling.

I give up Peterborough. His "Explanation" is neither straightforward, nor courteous, nor prudent. Of which last fact, it may be, he will be convinced when he reads my acknowledgment of his favours, which is soft, not with the softness of the answer which turneth away wrath, but with that of the pillow which smothered Desdemona.—Ever yours very faithfully,

> T. H. HUXLEY.

I shall try to stand an hour or two of the Academy dinner, and hope it won't knock me up.

---

[1] *I.e.* in the French sense of the word.

4 MARLBOROUGH PLACE, N.W.,
May 6, 1889.

MY DEAR KNOWLES—If I had not gone to the Academy dinner I might have kept my promise about sending you my paper to-day. I indulged in no gastronomic indiscretions, and came away after H.R.H.'s speech, but I was dead beat all yesterday, nevertheless.

We are off to Eastbourne, and I will send the MS. from there ; there is very little to do.

Such a waste ! I shall have to omit a paragraph that was really a masterpiece.

For who should I come upon in one of the rooms but the Bishop ! As we shook hands, he asked whether that was before the fight or after ; and I answered, " A little of both." Then we spoke our minds pretty plainly ; and then we agreed to bury the hatchet.[1]

So yesterday I tore up *the* paragraph. It was so appropriate I could not even save it up for somebody else !—Ever yours,   T. H. HUXLEY.

3 JEVINGTON GARDENS, EASTBOURNE,
May 22, 1889.

MY DEAR KNOWLES—I sent back my proof last evening. I shall be in town Friday afternoon to Monday morning next, having a lot of things to do. So you may as well let me see a revise of the whole. Did you not say to me, " sitting by a sea-coal fire " (I say nothing about a " parcel gilt goblet "), that this screed was to be the " last word " ? I don't mind how long it goes on so long as I have the last word. But you must expect nothing from me for the next three or four months. We shall be off abroad, not later than the 8th June, and

---

[1] As he says (*Coll. Ess.* v. 210), this chance meeting ended " a temporary misunderstanding with a man of rare ability, candour, and wit, for whom I entertained a great liking and no less respect."

among the everlasting hills, a fico for your controversies! Wace's paper shall be waste paper for me. Oh! This is a "goak" which Peterborough would not understand.

I think you are right about the wine and water business —I had my doubts—but it was too tempting. All the teetotallers would have been on my side.

There is no more curious example of the influence of education than the respect with which this poor bit of conjuring is regarded. Your genuine pietist would find a mystical sense in thimblerig. I trust you have properly enjoyed the extracts from Newman. That a man of his intellect should be brought down to the utterance of such drivel—by Papistry, is one of the strongest of arguments against that damnable perverter of mankind, I know of. —Ever yours very faithfully,

T. H. HUXLEY.

Shortly afterwards, he received a long and rambling letter in connection with this subject. Referring to the passage in the first article, "the apostolic injunction to 'suffer fools gladly' should be the rule of life of a true agnostic," the writer began by begging him " to 'suffer gladly' one fool more," and after several pages wound up with a variation of the same phrase. It being impossible to give any valid answer to his hypothetical inquiries, Huxley could not resist the temptation to take the opening thus offered him, and replied :—

SIR—I beg leave to acknowledge your letter. I have complied with the request preferred in its opening paragraph.—Faithfully yours,   T. H. HUXLEY.

The following letter also arises out of this controversy :—

Its occasion (writes Mr. Taylor) was one which I had written on seeing an article in which he referred to the Persian sect of the Bâbis. I had read with much interest the account of it in Count Gobineau's book, and was much struck with the points of likeness to the foundation of Christianity, and the contrast between the subsequent history of the two ; I asked myself how, given the points of similarity, to account for the contrast ; is it due to the Divine within the one, or the human surroundings ? This question I put to Professor Huxley, with many apologies for intruding on his leisure, and a special request that he would not suffer himself to be further troubled by any reply.

To Mr. Robert Taylor

4 Marlborough Place, N.W.,
June 3, 1889.

Sir—In looking through a mass of papers, before I leave England for some months among the mountains in search of health, I have come upon your letter of 7th March. As a rule I find that out of the innumerable letters addressed to me, the only ones I wish to answer are those the writers of which are considerate enough to ask that they may receive no reply, and yours is no exception.

The question you put is very much to the purpose : a proper and full answer would take up many pages ; but it will suffice to furnish the heads to be filled up by your own knowledge.

1. The Church founded by Jesus has *not* made its way ; has *not* permeated the world—but *did* become extinct in the country of its birth—as Nazarenism and Ebionism.

2. The Church that did make its way and coalesced with the State in the 4th century had no more to do

with the Church founded by Jesus than Ultramontanism has with Quakerism. It is Alexandrian Judaism and Neoplatonistic mystagogy, and as much of the old idolatry and demonology as could be got in under new or old names.

3. Paul has said that the Law was schoolmaster to Christ with more truth than he knew. Throughout the Empire the synagogues had their cloud of Gentile hangers-on—those who "feared God"—and who were fully prepared to accept a Christianity which was merely an expurgated Judaism and the belief in Jesus as the Messiah.

4. The Christian "Sodalitia" were not merely religious bodies, but friendly societies, burial societies, and guilds. They hung together for all purposes—the mob hated them as it now hates the Jews in Eastern Europe, because they were more frugal, more industrious, and lived better lives than their neighbours, while they stuck together like Scotchmen.

If these things are so—and I appeal to your knowledge of history that they are so—what has the success of Christianity to do with the truth or falsehood of the story of Jesus?—I am, yours very faithfully,

T. H. HUXLEY.

The following letter was written in reply to one from Mr. Clodd on the first of the articles in this controversy. This article, it must be remembered, not only replied to Dr. Wace's attack, but at the same time bantered Mr. Frederic Harrison's pretensions on behalf of Positivism at the expense alike of Christianity and Agnosticism.

3 JEVINGTON GARDENS, EASTBOURNE,
Feb. 19, 1889.

MY DEAR MR. CLODD—I am very much obliged to you for your cheery and appreciative letter. If I do not

empty all Harrison's vials of wrath I shall be astonished !
But of all the sickening humbugs in the world, the sham
pietism of the Positivists is to me the most offensive.

I have long been wanting to say my say about these
questions, but my hands were too full. This time last
year I was so ill that I thought to myself, with Hamlet,
"the rest is silence." But my wiry constitution has
unexpectedly weathered the storm, and I have every
reason to believe that with renunciation of the devil
and all his works (*i.e.* public speaking, dining and being
dined, etc.) my faculties may be unimpaired for a good
spell yet. And whether my lease is long or short, I
mean to devote them to the work I began in the paper
on the Evolution of Theology.

You will see in the next *Nineteenth* a paper on the
Evidence of Miracles, which I think will be to your
mind.

Hutton is beginning to drivel.[1] There really is no
other word for it.—Ever yours very faithfully,

                                        T. H. HUXLEY.

TO THE SAME

                                4 MARLBOROUGH PLACE,
                                    April 15, 1889.

MY DEAR MR. CLODD — The adventurous Mr. C.
wrote to me some time ago. I expressed my regret that
I could do nothing for the evolution of tent-pegs. What
wonderful people there are in the world !

Many thanks for calling my attention to "Antiqua
Mater." I will look it up. I have such a rooted

---

[1] This refers to an article in the *Spectator* on "Professor
Huxley and Agnosticism," Feb. 9, 1889, which suggests, with
regard to demoniac possession, that the old doctrine of one spirit
driving out another is as good as any new explanation, and
fortifies this conclusion by a reference to the phenomena of
hypnotism.

objection to returning books, that I never borrow one or allow anybody to lend me one if I can help it.

I hear that Wace is to have another innings, and I am very glad of it, as it will give me the opportunity of putting the case once more as a connected argument.

It is Baur's great merit to have seen that the key to the problem of Christianity lies in the Epistle to the Galatians. No doubt he and his followers rather over-did the thing, but that is always the way with those who take up a new idea.

I have had for some time the notion of dealing with the "Three great myths"—1. Creation; 2. Fall; 3 Deluge; but I suspect I am getting to the end of my tether physically, and shall have to start for the Engadine in another month's time.

Many thanks for your congratulations about my daughter's marriage. No two people could be better suited for one another, and there is a charming little grand-daughter of the first marriage to be cared for.— Ever yours very faithfully,   T. H. HUXLEY.

One more piece of writing dates from this time. He writes to his wife on March 2 :—

A man who is bringing out a series of portraits of celebrities, with a sketch of their career attached, has bothered me out of my life for something to go with my portrait, and to escape the abominable bad taste of some of the notices, I have done that. I shall show it you before it goes back to Engel in proof.

This sketch of his life is the brief autobiography which is printed at the beginning of vol. i. of the *Collected Essays*. He was often pressed, both by friends and by strangers, to give them some more autobiography ; but moved either by dislike of any approach to egotism, or by the knowledge that if biography is liable to give a false impression, autobiography may leave one still more false, he constantly refused to do so, especially so long as he had capacity for useful work. I found, however, among his papers, an entirely different sketch of his early life, half-a-dozen sheets describing the time he spent in the East end, with an almost Carlylean sense of the horrible disproportions of life. I cannot tell whether this was a first draft for the present autobiography, or the beginnings of a larger undertaking.

Several letters of miscellaneous interest were written before the move to the Engadine took place. They touch on such points as the excessive growth of scientific clubs, the use of alcohol for brain workers, advice to one who was not likely to "suffer fools gladly" about applying for the assistant secretaryship of the British Association, and the question of the effects of the destruction of immature fish, besides personal matters.

3 JEVINGTON GARDENS, EASTBOURNE,
March 22, 1889.

MY DEAR HOOKER—I suppose the question of amalgamation with the Royal is to be discussed at the Phil. Club. The sooner something of the kind takes place the better. There is really no *raison d'être* left for the Phil. Club, and considering the hard work of scientific men in these days, clubs are like hypotheses, not to be multiplied beyond necessity.—Ever yours,

T. H. HUXLEY.

4 MARLBOROUGH PLACE,
March 26, 1889.

MY DEAR HOOKER—The only science to which X. has contributed, so far as I know, is the science of self-advertisement; and of that he is a master.

When you and I were youngsters, we thought it the great thing to exorcise the aristocratic flunkeyism which reigned in the R.S.—the danger now is that of the entry of seven devils worse than the first, in the shape of rich engineers, chemical traders, and "experts" (who have sold their souls for a good price), and who find it helps them to appear to the public as if they were men of science.

If the Phil. Club had kept pure, it might have acted as a check upon the intrusion of the mere trading element. But there seems to be no reason now against Jack and Tom and Harry getting in, and the thing has become an imposture.

So I go with you for extinction, before we begin to drag in the mud.

I wish I could take some more active part in what is going on. I am anxious about the Society altogether. But though I am wonderfully well so long as I live like a hermit, and get out into the air of the Downs, either London, or bother, and still more both combined, intimate respectfully but firmly, that my margin is of the narrowest. —Ever yours,   T. H. HUXLEY.

The following is to his daughter in Paris. Of course it was the Tuileries, not the Louvre, which was destroyed in 1871 :—

I think you are quite right about French women. They are like French dishes, uncommonly well cooked and sent up, but what the dickens they are made of is a mystery. Not but what all womenkind are mysteries, but there are mysteries of godliness and mysteries of iniquity.

Have you been to see the sculptures in the Louvre ?—dear me, I forgot the Louvre's fate. I wonder where the sculpture is ? I used to think it the best thing in the way of art in Paris. There was a youthful Bacchus who was the main support of my thesis as to the greater beauty of the male figure !

Probably I had better conclude.

To Mr. E. T. Collings (of Bolton)

4 Marlborough Place,
April 9, 1889.

Dear Sir—I understand that you ask me what I think about "alcohol as a stimulant to the brain in mental work " ?

Speaking for myself (and perhaps I may add for persons of my temperament), I can say, without hesitation, that I would just as soon take a dose of arsenic as I would of alcohol, under such circumstances. Indeed on the whole, I should think the arsenic safer, less likely to lead to physical and moral degradation. It would be better to die outright than to be alcoholised before death.

If a man cannot do brain work without stimulants of any kind, he had better turn to hand work—it is an indication on Nature's part that she did not mean him to be a head worker.

The circumstances of my life have led me to experience all sorts of conditions in regard to alcohol, from total abstinence to nearly the other end of the scale, and my clear conviction is the less the better, though I by no means feel called upon to forgo the comforting and cheering effect of a little.

But for no conceivable consideration would I use it to

whip up a tired or sluggish brain.   Indeed, for me there
is no working time so good as between breakfast and
lunch, when there is not a trace of alcohol in my composition.

4 MARLBOROUGH PLACE,
May 6, 1889.

MY DEAR HOOKER—I meant to have turned up at the
x on Thursday, but I was unwell and, moreover, worried
and bothered about Collier's illness at Venice, and await-
ing [answer to] telegram I sent there.   He has contrived
to get scarlatina, but I hope he will get safe through it,
as he seems to be going on well.   We were getting ready
to go out until we were reassured on that point.

I thought I would go to the Academy dinner on
Saturday, and that if I did not eat and drink and came
away early, I might venture.

It was pleasant enough to have a glimpse of the world,
the flesh (on the walls, nude !), and the devil (there were
several Bishops), but oh, dear ! how done I was yesterday.

However, we are off to Eastbourne to-day, and I hope
to wash three weeks' London out of me before long.   I
think we shall go to Maloja again early in June.—Ever
yours,   T. H. HUXLEY.

Capital portrait in the New Gallery, where I looked
in for a quarter of an hour on Saturday—only you never
were quite so fat in the cheeks, and I don't believe you
have got such a splendid fur-coat !

3 JEVINGTON GARDENS, EASTBOURNE,
May 22, 1889.

. . . As to the Assistant Secretaryship of the British
Association, I have turned it over a great deal in my
mind since your letter reached me, and I really cannot
convince myself that you would suit it or it would suit
you.   I have not heard who are candidates or anything
about it, and I am not going to take any part in the

election. But looking at the thing solely from the point of view of your interests, I should strongly advise you against taking it, even if it were offered.

My pet aphorism "suffer fools gladly" should be the guide of the Assistant Secretary, who, during the fortnight of his activity, has more little vanities and rivalries to smooth over and conciliate than other people meet with in a lifetime. Now you do *not* "suffer fools gladly"; on the contrary, you "gladly make fools suffer." I do not say you are wrong—No *tu quoque*[1]—but that is where the danger of the explosion lies—not in regard to the larger business of the Association.

The risk is great and the £300 a year is not worth it. Foster knows all about the place; ask him if I am not right.

Many thanks for the suggestion about *Spirula*. But the matter is in a state in which no one can be of any use but myself. At present I am at the end of my tether and I mean to be off to the Engadine a fortnight hence— most likely not to return before October.

Not even the sweet voice of —— will lure me from my retirement. The Academy dinner knocked me up for three days, though I drank no wine, ate very little, and vanished after the Prince of Wales' speech. The truth is I have very little margin of strength to go upon even now, though I am marvellously better than I was.

I am very glad that you see the importance of doing battle with the clericals. I am astounded at the narrowness of view of many of our colleagues on this point. They shut their eyes to the obstacles which clericalism raises in every direction against scientific ways of thinking, which are even more important than scientific discoveries.

I desire that the next generation may be less fettered

---

[1] Cf. p. 114. But for due cause he could suffer them "with a difference"; of a certain caller he writes: "What an effusive bore he is! But I believe he was very kind to poor Clifford, and restrained my unregenerate impatience of that kind of creature."

by the gross and stupid superstitions of orthodoxy than mine has been. And I shall be well satisfied if I can succeed to however small an extent in bringing about that result.—I am, yours very faithfully,

T. H. HUXLEY.

4 MARLBOROUGH PLACE,
May 25, 1889.

MY DEAR LANKESTER—I cannot attend the Council meeting on the 29th. I have a meeting of the Trustees of the British Museum to-day, and to be examined by a Committee on Monday, and as the sudden heat half kills me I shall be fit for nothing but to slink off to Eastbourne again.

However, I do hope the Council will be very careful what they say or do about the immature fish question. The thing has been discussed over and over again *ad nauseam*, and I doubt if there is anything to be added to the evidence in the blue-books.

The *idée fixe* of the British public, fishermen, M.P.'s and ignorant persons generally is that all small fish, if you do not catch them, grow up into big fish. They cannot be got to understand that the wholesale destruction of the immature is the necessary part of the general order of things, from codfish to men.

You seem to have some very interesting things to talk about at the Royal Institution.

Do you see any chance of educating the white corpuscles of the human race to destroy the theological bacteria which are bred in parsons ?—Ever yours very faithfully,

T. H. HUXLEY.

3 JEVINGTON GARDENS, EASTBOURNE,
May 19, 1889.

MY DEAR DONNELLY—The Vice-President's letter has brought home to me one thing very clearly, and that is, that I had no business to sign the Report. Of course he has a right to hold me responsible for a document to which my name is attached, and I should look more like a fool than I ever wish to do, if I had to tell him that I had taken the thing entirely on trust. I have always objected to the sleeping partnership in the Examination ; and unless it can be made quite clear that I am nothing but a " consulting doctor," I really must get out of it entirely.

Of course I cannot say whether the Report is justified by the facts or not, when I do not know anything about them. But from my experience of what the state of things used to be, I should say that it is, in all probability, fair.

The faults mentioned are exactly those which always have made their appearance, and I expect always will do so, and I do not see why the attention of the teachers should not as constantly be directed to them. You talk of Eton. Well, the reports of the Examiners to the governing body, year after year, had the same unpleasing monotony, and I do not believe that there is any educational body, from the Universities downwards, which would come out much better, if the Examiners' reports were published and if they did their duty.

I am unable to see my way (and I suppose you are) to any better method of State encouragement of science teaching than payment by results. The great and manifest evil of that system, however, is the steady pressure which it exerts in the development of every description of sham teaching. And the only check upon this kind of swindling the public seems to me to lie in the hands of the Examiners. I told Mr. Forster so, ages ago, when he talked to me about the gradual increase of the expenditure, and I have been confirmed in my opinion by all subsequent experience. What the people who read the reports may say, I should not care one 2d. d— if I had to administer the thing.

Nine out of ten of them are incompetent to form any

opinion on an educational subject ; and as a mere matter of policy, I should, in dealing with them, be only too glad to be able to make it clear that some of the defects and shortcomings inherent in this (as in all systems) had been disguised, and that even the most fractious of Examiners had said their say without let or hindrance.

It is the nature of the system which seems to me to demand as a corrective incessant and severe watchfulness on the part of the Examiners, and I see no harm if they a little overdo the thing in this direction, for every sham they let through is an encouragement to other shams and pot-teaching in general.

And if the "great heart" of the people and its thick head can't be got to appreciate honesty, why the sooner we shut up the better.   Ireland may be for the Irish, but science teaching is not for the sake of science teachers.
—Ever yours,   T. H. HUXLEY.

CHAPTER VI

1889–90

FROM the middle of June to the middle of September, Huxley was in Switzerland, first at Monte Generoso, then, when the weather became more settled, at the Maloja. Here, as his letters show, he "rejuvenated" to such an extent that Sir Henry Thompson, who was at the Maloja, scoffed at the idea of his ever having had dilated heart.

MONTE GENEROSO, TESSIN, SUISSE,
June 25, 1889.

MY DEAR HOOKER—I am quite agreed with the proposed arrangements for the x, and hope I shall show better in the register of attendance next session.

When I am striding about the hills here I really feel as if my invalidism were a mere piece of malingering. When I am well I can walk up hill and down dale as well as I did twenty years ago. But my margin is abominably narrow, and I am at the mercy of "liver and lights." Sitting up for long and dining are questions of margin.

I do not know if you have been here. We are close on 4000 feet up and look straight over the great plain of N. Italy on the one side and to a great hemicycle of mountains, Monte Rosa among them, on the other. I do not know anything more beautiful in its way. But the whole time we have been here the weather has been extraordinary. On the average, about two thunderstorms *per diem.* I am sure that a good meteorologist might study the place with advantage. The barometer has not varied three-twentieths of an inch the whole time, notwithstanding the storms.

I hear the weather has been bad all over Switzerland, but it is not high and dry enough for me here, and we shall be off to the Maloja on Saturday next, and shall stay there till we return somewhere in September. Collier and Ethel will join us there in August. He is none the worse for his scarlatina.

"Aged Botanist?" marry come up ![1] I should like to know of a younger spark. The first time I heard myself called "the old gentleman" was years ago when we were in South Devon. A half-drunken Devonian had made himself very offensive, in the compartment in which my wife and I were travelling, and got some "simple Saxon" from me, accompanied, I doubt not, by an awful scowl. "Ain't the old gentleman in a rage," says he.

I am very glad to hear of Reggie's success, and my wife joins with me in congratulations. It is a comfort to see one's shoots planted out and taking root, though the idea that one's cares and anxieties about them are diminished, we find to be an illusion.

I inclose cheque for my contributions due and to come.[2] If I go to Davy's Locker before October, the latter may go for consolation champagne !—Ever yours affectionately,

T. H. HUXLEY.

He writes from the Maloja on August 16 to Sir

---

[1] Sir J. Hooker jestingly congratulated him on taking up botany in his old age.   [2] For the *x* Club.

M. Foster, who had been sitting on the Vaccination Commission :—

I wonder how you are prospering, whether you have vaccination or anti-vaccination on the brain ; or whether the gods have prospered you so far as to send you on a holiday. We have been here since the beginning of July. Monte Generoso proved lovely—but electrical. We had on the average three thunderstorms every two days. Bellagio was as hot as the tropics, and we stayed only a day, and came on here—where, whatever else may happen, it is never too hot. The weather has been good and I have profited immensely, and at present I do not know whether I have a heart or not. But I have to look very sharp after my liver. H. Thompson, who has been here with his son Herbert (clever fellow, by the way), treats the notion that I ever had a dilated heart with scorn ! Oh these doctors ! they are worse than theologians.

And again on August 31 :—

I walked eighteen miles three or four days ago, and I think nothing of one or two thousand feet up ! I hope this state of things will last at the sea-level.

I am always glad to hear of and from you, but I have not been idle long enough to forget what being busy means, so don't let your conscience worry you about answering my letters.

. . . X. is, I am afraid, more or less of an ass. The opposition he and his friends have been making to the Technical Bill is quite unintelligible to me. Y. may be, and I rather think is, a knave, but he is no fool ; and if I mistake not he is minded to kick the ultra-radical stool down now that he has mounted by it. Make friends of that Mammon of unrighteousness and swamp the sentimentalists.

. . . I despise your insinuations. All my friends here have been theological—Bishop, Chief Rabbi, and Catholic Professor. None of your Maybrick discussors.

On June 25 he wrote to Professor Ray Lankester, enclosing a letter to be read at a meeting called by the Lord Mayor, on July 1, to hear statements from men of science with regard to the recent increase of rabies in this country, and the efficiency of the treatment discovered by M. Pasteur for the prevention of hydrophobia.

I quote the latter from the report in *Nature* for July 4 :—

MONTE GENEROSO, TESSIN, SUISSE,
June 25, 1889.

MY DEAR LANKESTER—I enclose herewith a letter for the Lord Mayor and a cheque for £5 as my subscription.   I wish I could make the letter shorter, but it is pretty much "pemmican" already.   However, it does not much matter being read if it only gets into print.

It is uncommonly good of the Lord Mayor to stand up for Science, in the teeth of the row the anti-vivisection pack—dogs and doggesses—are making.

May his shadow never be less.

We shall be off to the Maloja at the end of this week, if the weather mends.   Thunderstorms here every day, and sometimes two or three a day for the last ten days.— Ever yours very faithfully,   T. H. HUXLEY.

MONTE GENEROSO, SWITZERLAND,
June 25, 1889.

MY LORD MAYOR—I greatly regret my inability to be present at the meeting which is to be held, under your Lordship's auspices, in reference to M. Pasteur and his Institute.   The unremitting labours of that eminent Frenchman during the last half-century have yielded rich harvests of new truths, and are models of exact and refined research.   As such they deserve, and have received, all the honours which those who are the best judges of their purely scientific merits are able to bestow. But it so happens that these subtle and patient searchings out of the ways of the infinitely little—of the swarming life where the creature that measures one-thousandth part of an inch is a giant—have also yielded results of supreme practical importance. The path of M. Pasteur's investigations is strewed with gifts of vast monetary value to the silk trades, the brewer, and the wine merchant. And this being so, it might well be a proper and graceful act on the part of the representatives of trade and commerce in its greatest centre to make some public recognition of M. Pasteur's services, even if there were nothing further to be said about them. But there is much more to be said. M. Pasteur's direct and indirect contributions to our knowledge of the causes of diseased states, and of the means of preventing their recurrence, are not measurable by money values, but by those of healthy life and diminished suffering to men. Medicine, surgery, and hygiene have all been powerfully affected by M. Pasteur's work, which has culminated in his method of treating hydrophobia. I cannot conceive that any competently instructed person can consider M. Pasteur's labours in this direction without arriving at the conclusion that, if any man has earned the praise and honour of his fellows, he has. I find it no less difficult to imagine that our wealthy country should be other than ashamed to continue to allow its citizens to profit by the treatment freely given at the Institute without contributing to its support. Opposition to the proposals which your Lordship sanctions would be equally inconceivable if it arose out of nothing but the facts of the case thus presented. But the opposition which, as I see from the English papers, is threatened has really for the most part nothing to do either with M. Pasteur's merits or with the efficacy of his method of treating hydrophobia. It proceeds partly from the fanatics of *laissez faire*, who think it better to

rot and die than to be kept whole and lively by State
interference, partly from the blind opponents of properly
conducted physiological experimentation, who prefer that
men should suffer than rabbits or dogs, and partly from
those who for other but not less powerful motives hate
everything which contributes to prove the value of strictly
scientific methods of enquiry in all those questions which
affect the welfare of society.   I sincerely trust that the
good sense of the meeting over which your Lordship will
preside will preserve it from being influenced by those
unworthy antagonisms, and that the just and benevolent
enterprise you have undertaken may have a happy issue.
—I am, my Lord Mayor, your obedient servant,

THOMAS H. HUXLEY.

HOTEL KURSAAL, MALOJA, HAUTE ENGADINE,
July 8, 1889.

MY DEAR LANKESTER—Many thanks for your letter.
I was rather anxious as to the result of the meeting,
knowing the malice and subtlety of the Philistines, but
as it turned out they were effectually snubbed.   I was
glad to see your allusion to Coleridge's impertinences.
It will teach him to think twice before he abuses his
position again.   I do not understand Stead's position in
the *Pall Mall.*   He snarls but does not bite.

I am glad that the audience (I judge from the *Times*
report) seemed to take the points of my letter, and live
in hope that when I see last week's *Spectator* I shall find
Hutton frantic.

This morning a letter marked " Immediate " reached
me from Bourne, date July 3.   I am afraid he does not
read the papers or he would have known it was of no
use to appeal to me in an emergency.   I am writing to
him.—Ever yours very faithfully,   T. H. HUXLEY.

On his return to England, however, a fortnight of
London, interrupted though it was by a brief visit to

Mr. and Mrs. Humphry Ward at the delightful old house of Great Hampden, was as much as he could stand. "I begin to discover," he writes to Sir M. Foster, "I have a heart again, a circumstance of which I had no reminder at the Maloja." So he retreated at once to Eastbourne, which had done him so much good before.

> 4 MARLBOROUGH PLACE,
> *Sept.* 24, 1889.

MY DEAR HOOKER—How's a' wi' ye? We came back from the Engadine early in the month, and are off to Eastbourne to-morrow. I rejuvenate in Switzerland and senescate (if there is no such verb, there ought to be) in London, and the sooner I am out of it the better.

When are you going to have an *x*? I cannot make out what has become of Spencer, except that he is some-where in Scotland.—Ever yours,   T. H. HUXLEY.

We shall be at our old quarters—3 Jevington Gardens, Eastbourne—from to-morrow onwards.

The next letter shows once more the value he set upon botanical evidence in the question of the influence of conditions in the process of evolution.

> 3 JEVINGTON GARDENS, EASTBOURNE,
> *Sept.* 29, 1889.

MY DEAR HOOKER—I hope to be with you at the Athenæum on Thursday. It does one good to hear of your being in such good working order. My knowledge of orchids is infinitesimally small, but there were some eight or nine species plentiful in the Engadine, and I learned enough to appreciate the difficulties. Why do not some of these people who talk about the direct influence of conditions try to explain the structure of orchids on that tack ? Orchids at any rate can't try to improve themselves in taking shots at insects' heads with pollen bags—as Lamarck's Giraffes tried to stretch their necks !

Balfour's *ballon d'essai* [1] (I do not believe it could have been anything more) is the only big blunder he has made, and it passes my comprehension why he should have made it. But he seems to have dropped it again like the proverbial hot potato. If he had not, he would have hopelessly destroyed the Unionist party.—Ever yours,

T. H. HUXLEY.

At the end of the year he thanks Lord Tennyson for his gift of " Demeter."

Dec. 26, 1889.

MY DEAR TENNYSON—Accept my best thanks for your very kind present of " Demeter." I have not had a Christmas Box I valued so much for many a long year. I envy your vigour, and am ashamed of myself beside you for being turned out to grass. I kick up my heels now and then, and have a gallop round the paddock, but it does not come to much.

With best wishes to you, and, if Lady Tennyson has not forgotten me altogether, to her also—Believe me, yours very faithfully,   T. H. HUXLEY.

A discussion in the *Times* this autumn, in which he joined, was of unexpected moment to him, inasmuch as it was the starting-point for no fewer than four essays in political philosophy, which appeared the following year in the *Nineteenth Century.*

The correspondence referred to arose out of the heckling of Mr. John Morley by one of his constituents at Newcastle in November 1889. The heckler

---

[1] *I.e.* touching a proposed Roman Catholic University for Ireland.

questioned him concerning private property in land, quoting some early dicta from the "'Social Statics" of Mr. Herbert Spencer, which denied the justice of such ownership. Comments and explanations ensued in the *Times*; Mr. Spencer declared that he had since partly altered that view, showing that contract has in part superseded force as the ground of ownership; and that in any case it referred to the idea of absolute ethics, and not to relative or practical politics.

Huxley entered first into the correspondence to point out present and perilous applications of the absolute in contemporary politics. Touching on a State guarantee of the title to land, he asks if there is any moral right for confiscation :—In Ireland, he says, confiscation is justified by the appeal to wrongs inflicted a century ago; in England the theorems of "absolute political ethics" are in danger of being employed to make this generation of land-owners responsible for the misdeeds of William the Conqueror and his followers. (*Times*, November 12.)

His remaining share in the discussion consisted of a brief passage of arms with Mr. Spencer on the main question,[1] and a reply to another correspondent,[2] which brings forward an argument enlarged upon in one of the essays, viz. that if the land belongs to all men equally, why should one nation claim one portion rather than another ? For several ownership is just as much an infringement of the world's ownership as is

[1] November 18.   [2] November 21.

personal ownership. Moreover, history shows that land was originally held in several ownership, and that not of the nation, but of the village community.

These signs of renewed vigour induced Mr. Knowles to write him a "begging letter," proposing an article for the *Nineteenth Century* either in commendation of Bishop Magee's recent utterances—it would be fine for eulogy to come from such a quarter after the recent encounter—or on the general subject of which his *Times* letters dealt with a part.

Huxley's choice was for the latter. Writing on November 21, he says :—

> Now as to the article. I have only hesitated because I want to get out a new volume of essays, and I am writing an introduction which gives me an immensity of trouble. I had made up my mind to get it done by Christmas, and if I write for you it won't be. However, if you don't mind leaving it open till the end of this month, I will see what can be done in the way of a screed about, say, "The Absolute in Practical Life." The Bishop would come in excellently ; he deserves all praises, and my only hesitation about singing them is that the conjunction between the "Infidel" and the Churchman is just what the blatant platform Dissenters who had been at him would like. I don't want to serve the Bishop, for whom I have a great liking and respect, as the bear served his sleeping master, when he smashed his nose in driving an unfortunate fly away !
>
> By the way, has the Bishop published his speech or sermon ? I have only seen a newspaper report.

Soon after this, he proposed to come to town and talk over the article with Mr. Knowles. The latter sent him a telegram—reply paid—asking him to fix a day. The answer named a day of the week and a day of the month which did not agree; whereupon Mr. Knowles wrote by the safer medium of the post for an explanation, thinking that the post-office clerks must have bungled the message, and received the following reply :—

> 3 JEVINGTON GARDENS, EASTBOURNE,
> *Nov.* 26, 1889.
>
> MY DEAR KNOWLES—May jackasses sit upon the graves of all telegraph clerks! But the boys are worse, and I shall have to write to the P.-M.-General about the little wretch who brought your telegram the other day, when my mind was deeply absorbed in the concoction of an article for *the* Review of our age.
>
> The creature read my answer, for he made me pay three halfpence extra (I believe he spent it on toffy), and yet was so stupid as not to see that meaning to fix next Monday or Tuesday, I opened my diary to give the dates in order that there should be no mistake, and found Monday 28 and Tuesday 29.
>
> And I suppose the little beast would say he did not know I opened it in October instead of November!
>
> I hate such mean ways. Hang all telegraph boys!—
> Ever yours very faithfully,   T. H. HUXLEY.
>
> Monday December 2, if you have nothing against it, and lunch if Mrs. Knowles will give me some.

The article was finished by the middle of December and duly sent to the editor, under the title of "Rousseau and Rousseauism." But fearing that this title would scarcely attract attention among the working-men for whom it was specially designed, Mr.

Knowles suggested instead the "Natural Inequality of Men," under which name it actually appeared in January. So, too, in the case of a companion article in March, the editorial pen was responsible for the change from the arid possibilities of "Capital and Labour" to the more attractive title of "Capital the Mother of Labour."

With regard to this article and a further project of extending his discussion of the subject, he writes :—

3 JEVINGTON GARDENS, EASTBOURNE,
Dec. 14, 1889.

MY DEAR KNOWLES—I am very glad you think the article will go. It is longer than I intended, but I cannot accuse myself of having wasted words, and I have left out several things that might have been said, but which can come in by and by.

As to title, do as you like, but that you propose does not seem to me quite to hit the mark. "Political Humbug : Liberty and Equality," struck me as adequate, but my wife declares it is improper. "'Political Fictions" might be supposed to refer to Dizzie's novels ! How about "The Politics of the Imagination : Liberty and Inequality " ?

I should like to have some general title that would do for the "letters" which I see I shall have to write. I think I will make six of them after the fashion of my "Working Men's Lectures," as thus : (1) Liberty and Equality ; (2) Rights of Man ; (3) Property ; (4) Malthus ; (5) Government, the province of the State ; (6) Law-making and Law-breaking.

I understand you will let me republish them, as soon as the last is out, in a cheap form. I am not sure I will not put them in the form of "Lectures" rather than "Letters."

Did you ever read Henry George's book "Progress and Poverty"? It is more damneder nonsense than poor Rousseau's blether. And to think of the popularity of the book! But I ought to be grateful, as I can cut and come again at this wonderful dish.

The mischief of it is I do not see how I am to finish the introduction to my Essays, unless I put off sending you a second dose until March.

I will send back the revise as quickly as possible.—
Ever yours very truly,   T. H. HUXLEY.

You do not tell me that there is anything to which Spencer can object, so I suppose there is nothing.

And in an undated letter to Sir J. Hooker, he says :—

I am glad you think well of the "Human Inequality" paper. My wife has persuaded me to follow it up with a view to making a sort of "Primer of Politics" for the masses—by and by. "There's no telling what you may come to, my boy," said the Bishop who reproved his son for staring at John Kemble, and I may be a pamphleteer yet! But really it is time that somebody should treat the people to common sense.

However, immediately after the appearance of this first article on Human Inequality, he changed his mind about the Letters to Working Men, and resolved to continue what he had to say in the form of essays in the *Nineteenth Century*.

He then judged it not unprofitable to call public attention to the fallacies which first found their way into practical politics through the disciples of Rousseau; one of those speculators of whom he remarks (*Coll. Ess.* i. 312) that "busied with deduction from their ideal 'ought to be,' they overlooked the 'what has been,' the 'what is,' and the 'what can be.'" "Many a long year ago," he says in *Natural Rights and Political Rights* (i. 336), "I fondly imagined that Hume and Kant and Hamilton having slain the 'Absolute,' the thing must, in decency, decease. Yet, at the present time, the same hypostatised negation, sometimes thinly disguised under a new name, goes about in broad daylight, in company with the dogmas of absolute ethics, political and other, and seems to be as lively as ever." This was to his mind one of those instances of wrong thinking which lead to wrong acting—the postulating a general principle based upon insufficient data, and the deduction from it of many and far-reaching practical consequences. This he had always strongly opposed. His essay of 1871, "Administrative Nihilism," was directed against *a priori* individualism; and now he proceeded to restate the arguments against *a priori* political reasoning in general, which seemed to have been forgotten or overlooked, especially by the advocates of compulsory socialism. And here it is possible to show in some detail the care he took, as was his way, to refresh his knowledge and bring it up to date, before writing on any special point. It is interesting to see how thoroughly he went to work, even in a subject with which he was already fairly acquainted. As in the controversy of 1889 I find a list of near a score of books consulted, so here one note-book contains an analysis of the origin and

early course of the French Revolution, especially in
relation to the speculations of the theorists; the
declaration of the rights of man in 1789 is followed
by parallels from Mably's *Droits et Devoirs du Citoyen*
and *De la Législation*, and by a full transcript of the
1793 Declaration, with notes on Robespierre's speech
at the Convention a fortnight later.   There are
copious notes from Dunoyer, who is quoted in the
article, while the references to Rocquain's *Esprit
Révolutionnaire* led to an English translation of the
work being undertaken, to which he contributed a
short preface in 1891.

It was the same with other studies.   He loved to
visualise his object clearly.   The framework of what
he wished to say would always be drawn out first.   In
any historical matter he always worked with a map.
In natural history he well knew the importance of
studying distribution and its bearing upon other
problems ; in civil history he would draw maps to
illustrate either the conditions of a period or the
spread of a civilising nation.   For instance, among
sketches of the sort which remain, I have one of the
Hellenic world, marked off in 25-mile circles from
Delos as centre ; and a similar one for the Phœnician
world, starting from Tyre.   Sketch maps of Palestine
and Mesopotamia, with notes from the best authorities
on the geography of the two countries, belong in all
probability to the articles on "The Flood" and
"Hasisadra's Adventure."   To realise clearly the size,
position, and relation of the parts to the whole, was

the mechanical instinct of the engineer which was so strong in him.

The four articles which followed in quick succession on " The Natural Inequality of Man," " Natural and Political Rights," "Capital the Mother of Labour," and "Government," appeared in the January, February, March, and May numbers of the *Nineteenth Century*, and, as was said above, are directed against *a priori* reasoning in social philosophy. The first, which appeared simultaneously with Mr. Herbert Spencer's article on "Justice," in the *Nineteenth Century*, assails, on the ground of fact and history, the dictum that men are born free and equal, and have a natural right to freedom and equality, so that property and political rights are a matter of contract. History denies that they thus originated ; and, in fact, " proclaim human equality as loudly as you like, Witless will serve his brother." Yet, in justice to Rousseau and the influence he wielded, he adds :—

It is not to be forgotten that what we call rational grounds for our beliefs are often extremely irrational attempts to justify our instincts.

Thus if, in their plain and obvious sense, the doctrines which Rousseau advanced are so easily upset, it is probable that he had in his mind something which is different from that sense.

When they sought speculative grounds to justify the empirical truth

that it is desirable in the interests of society, that all men should be as free as possible, consistently with those interests, and that they should all be equally bound by the ethical and legal obligations which are essential to social existence, "the philosophers," as is the fashion of speculators, scorned to remain on the safe if humble ground of experience, and preferred to prophesy from the sublime cloudland of the *a priori*.

The second of these articles is an examination of Henry George's doctrines as set forth in *Progress and Poverty*. His relation to the physiocrats is shown in a preliminary analysis of the term " natural rights which have no wrongs," and are antecedent to morality, from which analysis are drawn the results of confounding natural with moral rights.

Here again is the note of justice to an argument in an unsound shape (p. 369) : "There is no greater mistake than the hasty conclusion that opinions are worthless because they are badly argued." And a trifling abatement of the universal and exclusive form of Henry George's principle may make it true, while even unamended it may lead to opposite conclusions —to the justification of several ownership in land as well as in any other form of property.

The third essay of the series, "Capital the Mother of Labour" (*Coll. Ess.* ix. 147), was an application of biological methods to social problems, designed to show that the extreme claims of labour as against capital are ill-founded.

In the last article, "Government," he traces the two extreme developments of absolute ethics, as shown in anarchy and regimentation, or unrestrained individualism and compulsory socialism. The key to the position, of course, lies in the examination of the premisses upon which these superstructures are raised, and history shows that—

So far from the preservation of liberty and property and the securing of equal rights being the chief and most conspicuous object aimed at by the archaic politics of which we know anything, it would be a good deal nearer the truth to say that they were federated absolute monarchies, the chief purpose of which was the maintenance of an established church for the worship of the family ancestors.

These articles stirred up critics of every sort and kind; socialists who denounced him as an individualist, land nationalisers who had not realised the difference between communal and national ownership, or men who denounced him as an arm-chair cynic, careless of the poor and ignorant of the meaning of labour. Mr. Spencer considered the chief attack to be directed against his position; the regimental socialists as against theirs, and

as an attempt to justify those who, content with the present, are opposed to all endeavours to bring about any fundamental change in our social arrangements (*ib.* p. 423).

So far from this, he continues :—

Those who have had the patience to follow me to the end will, I trust, have become aware that my aim has been altogether different. Even the best of modern civilisations appears to me to exhibit a condition of mankind which neither embodies any worthy ideal nor even

possesses the merit of stability. I do not hesitate to
express my opinion that, if there is no hope of a large
improvement of the condition of the greater part of the
human family ; if it is true that the increase of knowledge,
the winning of a greater dominion over Nature which is
its consequence, and the wealth which follows upon that
dominion, are to make no difference in the extent and the
intensity of Want, with its concomitant physical and
moral degradation, among the masses of the people, I
should hail the advent of some kindly comet, which
would sweep the whole affair away, as a desirable consummation. What profits it to the human Prometheus that
he has stolen the fire of heaven to be his servant, and that
the spirits of the earth and of the air obey him, if the
vulture of pauperism is eternally to tear his very vitals
and keep him on the brink of destruction ?

Assuredly, if I believed that any of the schemes hitherto
proposed for bringing about social amelioration were
likely to attain their end, I should think what remains
to me of life well spent in furthering it. But my interest
in these questions did not begin the day before yesterday ;
and, whether right or wrong, it is no hasty conclusion of
mine that we have small chance of doing rightly in this
matter (or indeed in any other) unless we think rightly.
Further, that we shall never think rightly in politics
until we have cleared our minds of delusions, and more
especially of the philosophical delusions which, as I have
endeavoured to show, have infested political thought for
centuries. My main purpose has been to contribute my
mite towards this essential preliminary operation. Ground
must be cleared and levelled before a building can be
properly commenced ; the labour of the navvy is as
necessary as that of the architect, however much less
honoured ; and it has been my humble endeavour to grub
up those old stumps of the *a priori* which stand in the
way of the very foundations of a sane political philosophy.

To those who think that questions of the kind I have

been discussing have merely an academic interest, let me suggest once more that a century ago Robespierre and St. Just proved that the way of answering them may have extremely practical consequences.

Without pretending to offer any offhand solution for so vast a problem, he suggests two points in conclusion. One, that in considering the matter we should proceed from the known to the unknown, and take warning from the results of either extreme in self-government or the government of a family ; the other, that the central point is "the fact that the natural order of things—the order, that is to say, as unmodified by human effort—does not tend to bring about what we understand as welfare." The population question has first to be faced.

The following letters cover the period up to the trip to the Canaries, already alluded to :—

3 JEVINGTON GARDENS, EASTBOURNE,
Jan. 6, 1890.

MY DEAR FOSTER—That capital photograph reached me just as we were going up to town (invited for the holidays by our parents), and I put it in my bag to remind me to write to you.   Need I say that I brought it back again without having had the grace to send a line of thanks ?   By way of making my peace, I have told the Fine Art Society to send you a copy of the engraving of my sweet self.   I have not had it framed—firstly, because it is a hideous nuisance to be obliged to hang a frame one may not like ; and secondly, because by possibility you might like some other portrait better, in which case, if you will tell me, I will send that other.   I should like you to have something by way of reminder of T. H. H.

When Harry [1] has done his work at Bart's at the end of March I am going to give him a run before he settles down to practice. Probably we shall go to the Canaries. I hear that the man who knows most about them is Dr. Guillemard, a Cambridge man. " Kennst ihn du wohl ? " Perhaps he might give me a wrinkle.

With our united best wishes to you all—Ever yours very faithfully, T. H. HUXLEY.

EASTBOURNE, *Jan.* 13, 1890.

MY DEAR HOOKER—. . . We missed you on the 2nd, though you were quite right not to come in that beastly weather.

My boy Harry has had a very sharp attack of influenza at Bartholomew's, and came down to us to convalesce a week ago, very much pulled down. I hope you will keep clear of it.

H.'s work at the hospital is over at the end of March, and before the influenza business I was going to give him a run for a month or six weeks before he settled down to practice. We shall go to the Canaries as soon in April as possible. Are you minded to take a look at Teneriffe ? Only $4\frac{1}{2}$ days' sea—good ships.—Ever yours affectionately,

T. H. HUXLEY.

However, Sir J. Hooker was unable to join "the excursion to the Isles of the Blest."

EASTBOURNE, *Jan.* 27, 1890.

MY DEAR FOSTER—People have been at me to publish my notice of Darwin in *P.R.S.* in a separate form.

If you have no objection, will you apply to the Council for me for the requisite permission ?

But if you *do* see any objection, I would rather not make the request.

---

[1] His younger son.

I think if I republish it I will add the *Times* article of 1859 to it. Omega and Alpha !

Hope you are flourishing. We shall be up for a few days next week.—Ever yours very faithfully,

T. H. HUXLEY.

EASTBOURNE, Jan. 31, 1890.

MY DEAR FOSTER—Mind you let me know what points you think want expanding in the Darwin obituary when we meet.

We go to town on Tuesday for a few days, and I will meet you anywhere or anywhen you like. Could you come and dine with us at 4 P.M. on Thursday ? If so, please let me know at once, that E. may kill the fatted calf.

Harry has been and gone and done it. We heard he had gone to Yorkshire, and were anxious, thinking that at the very least a relapse after his influenza (which he had sharply) had occurred.

But the complaint was one with more serious *sequelæ* still. Don't know the young lady, but the youth has a wise head on his shoulders, and though that did not prevent Solomon from overdoing the business, I have every faith in his choice.

Dr. Guillemard has kindly sent me a lot of valuable information ; but as I suggested to my boy yesterday, he may find Yorkshire air more wholesome than that of the Canaries, and it is ten to one we don't go after all.—Ever yours,

T. H. H.

To his Younger Son

EASTBOURNE, Jan. 30, 1890.

YOU DEAR OLD HUMBUG OF A BOY—Here we have been mourning over the relapse of influenza, which alone, as we said, could have torn you from your duties, and all

the while it was nothing but an attack of palpitation
such as young people are liable to and seem none the
worse for after all. We are as happy that you are happy
as you can be yourself, though from your letter that
seems saying a great deal. I am prepared to be the
young lady's slave; pray tell her that I am a model
father-in-law, with my love. (By the way, you might
mention her name; it is a miserable detail, I know, but
would be interesting.) Please add that she is humbly
solicited to grant leave of absence for the Teneriffe trip,
unless she thinks Northallerton air more invigorating.—
Ever your loving dad,   T. H. HUXLEY.

On April 3, accompanied by his son, he left
London on board the *Aorangi*. At Plymouth he had
time to meet his friend W. F. Collier, and to visit
the Zoological Station, while, "to my great satisfaction," he writes, "I received a revise (*i.e.* of 'Capital
the Mother of Labour') for the May *Nineteenth Century*
—from Knowles. They must have looked sharp at
the printing-office."

It did not take him long to recover his sea-legs, and
he thoroughly enjoyed even the rougher days when
the rolling of the ship was too much for other people.
The day before reaching Teneriffe he writes :—

I have not felt so well for a long time. I do nothing,
have a prodigious appetite, and Harry declares I am
getting fat in the face.

Santa Cruz was reached early on April 10, and in
the afternoon he proceeded to Laguna, which he
made his headquarters for a week. That day he
walked 10 miles, the next 15, and the third 20 in

the course of the day. He notes finding the characteristic Euphorbia and Heaths of the Canaries; notes, too, one or two visitations of dyspepsia from indigestible food. He writes from Laguna :—

From all that people with whom we meet tell me, I gather that the usual massive lies about health resorts pervade the accounts of Teneriffe. Santa Cruz would reduce me to jelly in a week, and I hear that Orotava is worse—stifling. Guimar, whither we go to-morrow, is warranted to be dry and everlasting sunshine. We shall see. One of the people staying in the house said they had rain there for a fortnight together. . . . I am all right now, and walked some 15 miles up hill and down dale to-day, and I am not more than comfortably tired. However, I am not going to try the peak. I find it cannot be done without a night out at a considerable height when the thermometer commonly goes down below freezing, and I am not going to run that risk for the chance of seeing even the famous shadows.

By some mischance, no letters from home reached him till the 26th, and he writes from Guimar on the 23rd :—

A lady who lives here told me yesterday that a postmistress at one place was in the habit of taking off the stamps and turning the letters on one side ! But that luckily is not a particular dodge with ours.

We drove over here on the 17th. It is a very picturesque place 1000 feet up in the midst of a great amphitheatre of high hills, facing north, orange-trees laden with fruit, date palms and bananas are in the garden, and there is lovely sunshine all day long. Altogether the climate is far the best I have found anywhere here, and the house, which is that of a Spanish

Marquesa, only opened as a hotel this winter, is very comfortable. I am sitting with the window wide open at nine o'clock at night, and the stars flash as if the sky were Australian.

On Saturday we had a splendid excursion up to the top of the pass that leads from here up to the other side of the island. Road in the proper sense there was none, and the track incredibly bad, worse than any Alpine path owing to the loose irregular stones. The mules, however, pick their way like cats, and you have only to hold on. The pass is 6000 feet high, and we ascended still higher. Fortune favoured us. It was a lovely day and the clouds lay in a great sheet a thousand feet below. The peak, clear in the blue sky, rose up bare and majestic 5000 feet out of as desolate a desert [1] clothed with the stiff retama shrubs (a sort of broom) as you can well imagine. It took us three hours and a half to get up, passing for a good deal of the time through a kind of low brush of white and red cistuses in full bloom. We saw Palma on one side, and Grand Canary on the other, beyond the layer of clouds which enveloped all the lower part of the island. Coming down was worse than going up, and we walked a good part of the way, getting back about six. About seven hours in the saddle and walking.

You never saw anything like the improvement in Harry. He is burnt deep red ; he says my nose is of the same hue, and at the end of the journey he raced Gurilio, our guide, who understands no word of English any more than we do Spanish, but we are quite intimate nevertheless.[2]

---

[1] The Cañadas, which he calls "the one thing worth seeing there."

[2] My brother indeed averred that his language of signs was far more effectual than the Spanish which my father persisted in trying upon the inhabitants. This guide, by the way, was very sceptical as to any Englishman being equal to walking the seventeen miles, much less beating him in a race over the stony track. His experience was entirely limited to invalids.

He reiterates his distress at not getting letters from his wife : " Certainly I will never run the risk of being so long without—never again." When, after all, the delayed letters reached him on his way back from the expedition to the Canadas, thanks to a traveller who brought them up from Laguna, he writes (April 24) :—

Catch me going out of reach of letters again. I have been horridly anxious. Nobody—children or any one else—can be to me what you are. Ulysses preferred his old woman to immortality, and this absence has led me to see that he was as wise in that as in other things. . . .

Here is a novel description of an hotel at Puerto Orotava :—

It is very pretty to look at, but all draughts. I compare it to the air of a big wash-house with all the doors open, and it was agreed that the likeness was exact.

On May 2 he sailed for Madeira by the *German*, feeling already "ten years younger" for his holiday. On the 3rd he writes :—

The last time I was in this place was in 1846. All my life lies between the two visits. I was then twenty-one and a half, and I shall be sixty-five to-morrow. The place looks to me to have grown a good deal, but I believe it is chiefly English residents whose villas dot the hill. There were no roads forty-four years ago. Now there is one, I am told, to Camera do Lobos nearly five miles long. That is the measure of Portuguese progress in half a century. Moreover, the men have left off wearing their pigtail caps and the women their hoods.

To his Youngest Daughter

BELLA VISTA HOTEL, FUNCHAL,
May 6, 1890.

DEAREST BABS—This comes wishing you many happy
returns of the day, though a little late in the arrival.
Harry sends his love, and desires me to say that he took
care to write a letter which should arrive in time, but
unfortunately forgot to mention the birthday in it ! So
I think, on the whole, I have the pull of him.   We
ought to be back about the 18th or 19th, as I have put
my name down for places in the *Conway Castle*, which is
to call here on the 12th, and I do not suppose she will
be full.   In the meanwhile, we shall fill up the time by
a trip to the other side of the island, on which we start
to-morrow morning at 7.30.   You have to take your
own provisions and rugs to sleep upon and under, as the
fleas *la bas* are said to be unusually fine and active.   We
start quite a procession with a couple of horses, a guide,
and two men (owners of the nags) to carry the baggage ;
and I suspect that before to-morrow night we shall have
made acquaintance with some remarkably bad apologies
for roads.   But the horses here seem to prefer going up
bad staircases at speed (with a man hanging on by the
tail to steer), and if you only stick to them they land
you all right.   I have developed so much prowess in this
line that I think of coming out in the character of
Buffalo Bill on my return.   Hands and face of both of
us are done to a good burnt sienna, and a few hours more
or less in the saddle don't count.   I do not think either
of us have been so well for years.

   You will have heard of our doings in Teneriffe from
M———.   The Cañadas there is the one thing worth seeing, altogether unique.   As a health resort I should say
the place is a fraud—always excepting Guimar—and
that, excellent for people in good health, is wholly unfit

for a real invalid, who must either go uphill or downhill over the worst of roads if he leaves the hotel.

The air here is like that of South Devon at its best— very soft, but not stifling as at Orotava.   We had a capital expedition yesterday to the Grand Corral—the ancient volcanic crater in the middle of the island with walls some 3000 feet high, all scarred and furrowed by ravines, and overgrown with rich vegetation.   There is a little village at the bottom of it which I should esteem as a retreat if I wished to be out of sight and hearing of the pomps and vanities of this world.   By the way, I have been pretty well out of hearing of everything as it is, for I only had three letters from M—— while we were in Teneriffe, and not one here up to this date.   After I had made all my arrangements to start to-morrow I heard that a mail would be in at noon.   So the letters will have to follow us in the afternoon by one of the men, who will wait for them.

We went to-day to lunch with Mr. Blandy, the head of the principal shipping agency here, whose wife is the daughter of my successor at the Fishery Office.

Well, our trip has done us both a world of good ; but I am getting homesick, and shall rejoice to be back again. I hope that Joyce is flourishing, and Jack satisfied with the hanging of his pictures, and that a millionaire has insisted on buying the picture and adding a bonus.   Our best love to you all.—Ever your loving

PATER.

Don't know M——'s whereabouts.   But if she is with you, say I wrote her a long screed (No. 8) and posted it to-day—with my love as a model husband and complete letter-writer.

On returning home he found that the Linnean medal had been awarded him.

4 MARLBOROUGH PLACE,
May 18, 1890.

MY DEAR HOOKER—How's a' wi' you? My boy and
I came back from Madeira yesterday in great feather.
As for myself, riding about on mules, or horses, for six to
ten hours at a stretch—burning in sun or soaking in rain
-—over the most entirely breakneck roads and tracks I
have ever made acquaintance with, except perhaps in
Morocco—has proved a most excellent tonic, cathartic, and
alterative all in one. Existence of heart and stomach are
matters of faith, not of knowledge, with me at present.
I hope it may last, and I have had such a sickener of
invalidism that my intention is to keep severely out of
all imprudences.

But what is a man to do if his friends take advantage
of his absence, and go giving him gold medals behind his
back? That you have been an accomplice in this nefarious
plot—mine own familiar friend whom I trusted and
trust—is not to be denied. Well, it is very pleasant to
have toil that is now all ancient history remembered,
and I shall go to the meeting and the dinner and make
my speech in spite of as many possible devils of dyspepsia
as there are plates and dishes on the table.

We were lucky in getting in for nothing worse than
heavy rolling, either out or in. Teneriffe is well worth
seeing. The Canadas is something quite by itself, a bit
of Egypt 6000 feet up with a bare volcanic cone, or
rather long barrow sticking up 6000 feet in the middle
of it.

Otherwise, Madeira is vastly superior. I rode across
from Funchal to Sao Vicente, up to Paul da Serra, then
along the coast to Santa Anna, and back from Sta. Anna
to Funchal. I have seen nothing comparable except in
Mauritius, nor anything anywhere like the road by the
cliffs from Sao Vicente to Sta. Anna. Lucky for me
that my ancient nautical habit of sticking on to a horse

came back. A good deal of the road is like a bad stair-
case, with no particular banisters, and a well of 1000
feet with the sea at the bottom. Your heart would
rejoice over the great heaths. I saw one, the bole of which
split into nearly equal trunks; and one of these was
just a metre in circumference, and had a head as big as
a moderate-sized ash. Gorse in full flower, up to 12 or
15 feet high. On the whole a singular absence of flower-
ing herbs except *Cinerarias* and, especially in Teneriffe,
Echium. I did not chance to see a *Euphorbia* in Madeira,
though I believe there are some. In Teneriffe they are
everywhere in queer shapes, and there was a thing that
mimicked the commonest *Euphorbia* but had no milk,
which I will ask you about when I see you. The
Euphorbias were all in flower, but this thing had none.
But you will have had enough of my scrawl.—Ever yours
affectionately,   T. H. HUXLEY.

CHAPTER VII

1890–1891

THREE letters of the first half of the year may conveniently be placed here. The first is to Tyndall, who had just been delivering an anti-Gladstonian speech at Belfast. The opening reference must be to some newspaper paragraph which I have not been able to trace, just as the second is to a paragraph in 1876, not long after Tyndall's marriage, which described Huxley as starting for America with his titled bride.

3 JEVINGTON GARDENS, EASTBOURNE,
Feb. 24, 1890.

MY DEAR TYNDALL.—Put down the three half-pints and the two dozen to the partnership account. Ever since the "titled bride" business I have given up the struggle against the popular belief that you and I constitute a firm.

It's very hard on me in the decline of life to have a lively young partner who thinks nothing of rushing six or seven hundred miles to perform a war-dance on the sainted G.O.M., and takes the scalp of Historicus as a *hors d'œuvre.*

All of which doubtless goes down to my account just as my poor innocent articles confer a reputation for long-suffering mildness on you.

Well! well! there is no justice in this world! With our best love to you both—Ever yours, T. H. HUXLEY.

(The confusion in the popular mind continued steadily, so that at last, when Tyndall died, Huxley received the doubtful honour of a funeral sermon.)

Dr. Pelseneer, to whom the next letter is addressed, is a Belgian morphologist, and an authority upon the Mollusca. He it was who afterwards completed Huxley's unfinished memoir on Spirula for the *Challenger* report.

4 MARLBOROUGH PLACE,
June 10, 1890.

DEAR DR. PELSENEER—I gave directions yesterday for the packing up and sending to your address of the specimens of *Trigonia*, and I trust that they will reach you safely.

I am rejoiced that you are about to take up the subject. I was but a beginner when I worked at *Trigonia*, and I had always promised myself that I would try to make good the many deficiencies of my little sketch. But three or four years ago my health gave way completely, and though I have recovered (no less to my own astonishment than to that of the doctors) I am compelled to live out of London and to abstain from all work which involves much labour.

Thus science has got so far ahead of me that I hesitate to say much about a difficult morphological question—all the more, as old men like myself should be on their guard against over-much tenderness for their own speculations. And I am conscious of a great tenderness for those contained in my ancient memoir on the " Morphology of the Cephalous Mollusca." Certainly I am entirely disposed to agree with you that the Gasteropods and the Lamellibranchs spring from a common root — nearly represented by the Chiton—especially by a hypothetical *Chiton* with one shell plate.

I always thought *Nucula* the key to the Lamellibranchs, and I am very glad you have come to that conclusion on such much better evidence.—I am, dear Dr. Pelseneer, yours very faithfully,   T. H. HUXLEY.

Towards the end of June he went for a week to Salisbury, taking long walks in the neighbourhood, and exploring the town and cathedral, which he confessed himself ashamed never to have seen before.

He characteristically fixes its date in his memory by noting that the main part of it was completed when Dante was a year old.

THE WHITE HART, SALISBURY,
June 22, 1890.

MY DEAR DONNELLY—Couldn't stand any more London, so bolted here yesterday morning, and here I shall probably stop for the next few days.

I have been trying any time the last thirty years to see Stonehenge, and this time I mean to do it. I should have gone to-day, but the weather was not promising, so I spent my Sunday morning in Old Sarum—that blessed old tumulus with nine (or was it eleven?) burgesses that used to send two members to Parliament when I was a child. Really you Radicals are of some use after all !

Poor old Smyth's [1] death is just what I expected,

---

[1] Warington Wilkinson Smyth (1817-1890), the geologist and mineralogist. In 1851 he was appointed Lecturer on Mining and

though I did not think the catastrophe was so imminent.

Peace be with him ; he never did justice to his very considerable abilities, but he was a good fellow and a fine old crusted Conservative.

I suppose it will be necessary to declare the vacancy and put somebody in his place before long.

I learned before I started that Smyth was to be buried in Cornwall, so there is no question of attending at his funeral.

I am the last of the original Jermyn Street gang left in the school now—Ultimus Romanorum !—Ever yours very faithfully,   T. H. HUXLEY.

This trip was taken by way of a holiday after the writing of an article, which appeared in the *Nineteenth Century* for July 1890. It was called "The Lights of the Church and the Light of Science," and may be considered as written in fulfilment of the plan spoken of in the letter to Mr. Clodd (p. 117). Its subject was the necessary dependence of Christian theology upon the historical accuracy of the Old Testament ; its occasion, the publication of a sermon in which, as a counterblast to *Lux Mundi*, Canon Liddon declared that accuracy to be sanctioned by the use made of the Old Testament by Jesus Christ, and bade his hearers close their ears against any suggestions impairing the credit of those Jewish Scriptures which have received the stamp of His Divine authority.

Mineralogy at the Royal School of Mines. After the lectureships were separated in 1881, he retained the former until his death. He was knighted in 1887.

Pointing out that, as in other branches of history, so here the historical accuracy of early tradition was abandoned even by conservative critics, who at all understood the nature of the problems involved, Huxley proceeded to examine the story of the Flood, and to show that the difficulties were little less in treating it—like the reconcilers—as a partial than as a universal deluge.   Then he discussed the origin of the story, and criticised the attempt of the essayist in *Lux Mundi* to treat this and similar stories as "types," which must be valueless if typical of no underlying reality.   These things are of moment in speculative thought, for if Adam be not an historical character, if the story of the Fall be but a type, the basis of Pauline theology is shaken; they are of moment practically, for it is the story of the Creation which is referred to in the "speech (Matt. xix. 5) unhappily famous for the legal oppression to which it has been wrongfully forced to lend itself" in the marriage laws.

In July 1890, Sir J. G. T. Sinclair wrote to him, calling his attention to a statement of Babbage's that after a certain point his famous calculating machine, contrary to all expectation, suddenly introduced a new principle of numeration into a series of numbers,[1] and asking what effect this phenomenon had upon

---

[1] Extract from Babbage's Ninth Bridgewater Treatise.

Babbage shows that a calculating machine can be constructed which, after working in a correct and orderly manner up to 100,000,000, then leaps, and instead of continuing the chain of numbers unbroken, goes at once to 100,010,002.   "The law which seemed at first to govern the series failed at the hundred million

the theory of Induction. Huxley replied as follows :—

GRAND HOTEL, EASTBOURNE,
July 21, 1890.

DEAR SIR—I knew Mr. Babbage, and am quite sure that he was not the man to say anything on the topic of calculating machines which he could not justify.

I do not see that what he says affects the philosophy of induction as rightly understood. No induction, however broad its basis, can confer certainty—in the strict sense of the word. The experience of the whole human race through innumerable years has shown that stones unsupported fall to the ground, but that does not make it certain that any day next week unsupported stones will not move the other way. All that it does justify is the very strong expectation, which hitherto has been invariably verified, that they will do just the contrary.

---

and second term. This term is larger than we expected by 10,000. The law thus changes—

100,000,001	100,100,005
100,010,002	100,150,006
100,030,003	100,210,007
100,060,004	100,280,008.

For a hundred or even a thousand terms they continued to follow the new law relating to the triangular numbers, but after watching them for 2761 terms we find that this law fails at the 2762nd term.

If we continue to observe we shall discover another law then coming into action which also is different, dependent, but in a different manner, on triangular numbers because a number of points agreeing with their term may be placed in the form of a triangle, thus—

.
. . .
. . . . . .
. . . . (one, three, six, ten).

This will continue through about 1430 terms, when a new law is again introduced over about 950 terms, and this too, like its predecessors, fails and gives place to other laws which appear at different intervals."

Only one absolute certainty is possible to man—namely, that at any given moment the feeling which he has exists.

All other so-called certainties are beliefs of greater or less intensity.

Do not suppose that I am following Abernethy's famous prescription, "take my pills," if I refer you to an essay of mine on "Descartes," and a little book on Hume, for the fuller discussion of these points. Hume's argument against miracles turns altogether on the fallacy that induction can give certainty in the strict sense.

We poor mortals have to be content with hope and belief in all matters past and present—our sole certainty is momentary.—I am yours faithfully,

T. H. HUXLEY.

Sir J. G. T. Sinclair, Bart.

Except for a last visit to London to pack his books, which proved a heavier undertaking than he had reckoned upon, Huxley did not leave Eastbourne this autumn, refusing Sir J. Donnelly's hospitable invitation to stay with him in Surrey during the move, of which he exclaims :—

Thank Heaven that is my last move—except to a still smaller residence of a subterranean character !

GRAND HOTEL, EASTBOURNE,
Sept. 19, 1890.

MY DEAR DONNELLY—And my books—and watch-dog business generally ?

How is that to be transacted whether as in-patient or out-patient at Firdale ? Much hospitality hath made thee mad.

Seriously, it's not to be done nohow. What between papers that don't come, and profligate bracket manu-

facturers who keep you waiting for months and then send
the wrong things—and a general tendency of everybody
to do nothing right or something wrong—it is as much
as the two of us will do—to get in, and all in the course
of the next three weeks.

Of course my wife has no business to go to London to
superintend the packing—but I should like to see anybody stop her.   However, she has got the faithful Minnie
to do the actual work ; and swears by all her Gods and
Goddesses she will only direct.

It would only make her unhappy if I did not make
pretend to believe, and hope no harm may come of it.—
Tout à vous,   T. H. HUXLEY.

Another discussion which sprang up in the *Times*,
upon Medical Education, evoked a letter from him
(*Times*, August 7), urging that the preliminary training ought to be much more thorough and exact.   The
student at his first coming is so completely habituated
to learn only from books or oral teaching, that the
attempt to learn from things and to get his knowledge
at first hand is something new and strange.   Thus a
large proportion of medical students spend much of
their first year in learning how to learn, and when
they have done that, in acquiring the preliminary
scientific knowledge, with which, under any rational
system of education, they would have come provided.

He urged, too, that they should have received a
proper literary education instead of a sham acquaintance with Latin, and insisted, as he had so often
done, on the literary wealth of their own language.

Every one has his own ideas of what a liberal

education ought to include, and a correspondent
wrote to ask him, among other things, whether he
did not think the higher mathematics ought to be
included. He replied :—

GRAND HOTEL, EASTBOURNE,
Aug. 16, 1890.

I think mathematical training highly desirable, but
advanced mathematics, I am afraid, would be too great
a burden in proportion to its utility, to the ordinary
student.

I fully agree with you that the incapacity of teachers
is the weak point in the London schools. But what is
to be expected when a man accepts a lectureship in a
medical school simply as a grappling-iron by which he
may hold on until he gets a hospital appointment ?

Medical education in London will never be what it
ought to be, until the " Institutes of Medicine," as the
Scotch call them, are taught in only two or three well-
found institutions—while the hospital schools are confined
to the teaching of practical medicine, surgery, obstetrics,
and so on.

The following letters illustrate Huxley's keenness
to correct any misrepresentation of his opinions from
a weighty source, and the way in which, without
abating his just claims, he could make the peace
gracefully.

In October Dr. Abbott delivered an address on
" Illusions," in which, without, of course, mentioning
names, he drew an unmistakable picture of Huxley as
a thorough pessimist. A very brief report appeared
in the *Times* of October 9, together with a leading
article upon the subject. Huxley thereupon wrote

to the *Times* a letter which throws light both upon his early days and his later opinions :—

The article on "Illusions" in the *Times* of to-day induces me to notice the remarkable exemplification of them to which you have drawn public attention. The Rev. Dr. Abbott has pointed the moral of his discourse by a reference to a living man, the delicacy of which will be widely and justly appreciated. I have reason to believe that I am acquainted with this person, somewhat intimately, though I can by no means call myself his best friend—far from it.

If I am right, I can affirm that this poor fellow did not escape from the "narrow school in which he was brought up" at nineteen, but more than two years later ; and, as he pursued his studies in London, perhaps he had as many opportunities for "fruitful converse with friends and equals," to say nothing of superiors, as he would have enjoyed elsewhere.

Moreover, whether the naval officers with whom he consorted were book-learned or not, they were emphatically men, trained to face realities and to have a wholesome contempt for mere talkers. Any one of them was worth a wilderness of phrase-crammed undergraduates. Indeed, I have heard my misguided acquaintance declare that he regards his four years' training under the hard conditions and the sharp discipline of his cruise as an education of inestimable value.

As to being a "keen-witted pessimist out and out," the Rev. Dr. Abbott's "horrid example" has shown me the following sentence :—" Pessimism is as little consonant with the facts of sentient existence as optimism." He says he published it in 1888, in an article on "Industrial Development," to be seen in the *Nineteenth Century.* But no doubt this is another illusion. No superior person, brought up "in the Universities," to boot, could possibly have invented a myth so circumstantial.

The end of the correspondence was quite amicable.
Dr. Abbott explained that he had taken his facts
from the recently published "Autobiography," and
that the reporters had wonderfully altered what he
really said by large omissions.　In a second letter
(*Times*, October 11) Huxley says :—

I am much obliged to Dr. Abbott for his courteous
explanation.　I myself have suffered so many things at
the hands of so many reporters—of whom it may too
often be said that their "faith, unfaithful, makes them
falsely true"—that I can fully enter into what his feelings
must have been when he contemplated the picture of his
discourse, in which the lights on "raw midshipmen,
"pessimist out and out," "devil take the hindmost," and
"Heine's dragoon," were so high, while the "good things"
he was kind enough to say about me lay in the deep
shadow of the invisible.　And I can assure Dr. Abbott
that I should not have dreamed of noticing the report of
his interesting lecture, which I read when it appeared,
had it not been made the subject of the leading article
which drew the attention of all the world to it on the
following day.

I was well aware that Dr. Abbott must have founded
his remarks on the brief notice of my life which (without
my knowledge) has been thrust into its present ridiculous
position among biographies of eminent musicians ; and most
undoubtedly anything I have said there is public property.
But erroneous suppositions imaginatively connected with
what I have said appear to me to stand upon a different
footing, especially when they are interspersed with
remarks injurious to my early friends.　Some of the
"raw midshipmen and unlearned naval officers" of whom
Dr. Abbott speaks, in terms which he certainly did not
find in my "autobiography," are, I am glad to say, still
alive, and are performing, or have performed, valuable

services to their country. I wonder what Dr. Abbott would think, and perhaps say, if his youthful University friends were spoken of as "raw curates and unlearned country squires."

When David Hume's housemaid was wroth because somebody chalked up "St. David's" on his house, the philosopher is said to have remarked,—"Never mind, lassie, better men than I have been made saints of before now." And, perhaps, if I had recollected that "better men than I have been made texts of before now," a slight flavour of wrath which may be perceptible would have vanished from my first letter. If Dr. Abbott has found any phrase of mine too strong, I beg him to set it against "out and out pessimist" and "Heine's dragoon," and let us cry quits. He is the last person with whom I should wish to quarrel.

Two interesting criticisms of books follow; one *The First Three Gospels*, by the Rev. Estlin Carpenter; the other on *Use and Disuse*, directed against the doctrine of use-inheritance, by Mr. Platt Ball, who not only sent the book but appealed to him for advice as to his future course in undertaking a larger work on the evolution of man.

GRAND HOTEL, EASTBOURNE,
Oct. 11, 1890.

MY DEAR MR. CARPENTER—Accept my best thanks for *The First Three Gospels*, which strikes me as an admirable exposition of the case, full, clear, and calm. Indeed the latter quality gives it here and there a touch of humour. You say the most damaging things in a way so gentle that the orthodox reader must feel like the eels who were skinned by the fair Molly—lost between pain and admiration

I am certainly glad to see that the book has reached a second edition ; it will do yeoman's service to the cause of right reason.

A friend of mine was in the habit of sending me his proofs, and I sometimes wrote on them "no objection except to the whole" ; and I am afraid that you will think what I am about to say comes to pretty much the same thing—at least if I am right in the supposition that a passage in your first preface (p. vii.) states your fundamental position, and that you conceive that when criticism has done its uttermost there still remains evidence that the personality of Jesus was the leading cause—the *conditio sine qua non*—of the evolution of Christianity from ʃudaism.

I long thought so, and having a strong dislike to belittle the heroic figures of history, I held by the notion as long as I could, but I find it melting away.

I cannot see that the moral and religious ideal of early Christianity is new—on the other hand, it seems to me to be implicitly and explicitly contained in the early prophetic Judaism and the later Hellenised Judaism ; and though it is quite true that the new vitality of the old ideal manifested in early Christianity demands "an adequate historic cause," I would suggest that the word "cause" may mislead if it is not carefully defined.

Medical philosophy draws a most useful and necessary distinction between "exciting" and "predisposing" causes —and nowhere is it more needful to keep this distinction in mind than in history—and especially in estimating the action of individuals on the course of human affairs. Platonic and Stoical philosophy—prophetic liberalism— the strong democratic socialism of the Jewish political system — the existence of innumerable sodalities for religious and social purposes—had thrown the ancient world into a state of unstable equilibrium. With such predisposing causes at work, the exciting cause of enormous changes might be relatively insignificant. The powder was there—a child might throw the match which should blow up the whole concern.

I do not want to seem irreverent, still less depreciatory, of noble men, but it strikes me that in the present case the Nazarenes were the match and Paul the child.

An ingrained habit of trying to explain the unknown by the known leads me to find the key to Nazarenism in Quakerism. It is impossible to read the early history of the Friends without seeing that George Fox was a person who exerted extraordinary influence over the men with whom he came in contact; and it is equally impossible (at least for me) to discover in his copious remains an original thought.

Yet what with the corruption of the Stuarts, the Phariseeism of the Puritans, and the Sadduceeism of the Church, England was in such a state, that before his death he had gathered about him a vast body of devoted followers, whose patient endurance of persecution is a marvel. Moreover, the Quakers have exercised a prodigious influence on later English life.

But I have scribbled a great deal too much already. You will see what I mean.

To Mr. W. Platt Ball

GRAND HOTEL, EASTBOURNE,
Oct. 27, 1890.

DEAR SIR—I have been through your book, which has greatly interested me, at a hand-gallop; and I have by no means given it the attention it deserves. But the day after to-morrow I shall be going into a new house here, and it may be some time before I settle down to work in it—so that I prefer to seem hasty, rather than indifferent to your book and still more to your letter.

As to the book, in the first place. The only criticism I have to offer—in the ordinary depreciatory sense of the word—is that pp. 128-137 seem to me to require reconsideration, partly from a substantial and partly from a tactical point of view. There is much that is disputable on the one hand, and not necessary to your argument on the other.

Otherwise it seems to me that the case could hardly be better stated. Here are a few notes and queries that have occurred to me.

P. 41. Extinction of Tasmanians—rather due to the British colonist, who was the main agent of their extirpation, I fancy.

P. 67. Birds' sternums are a great deal more than surfaces of origin for the pectoral muscles—*e.g.* movable lid of respiratory bellows. This not taken into account by Darwin.

P. 85. "Inferiority of senses of Europeans" is, I believe, a pure delusion. Prof. Marsh told me of feats of American trappers equal to any savage doings. It is a question of attention. Consider wool-sorters, tea-tasters, shepherds who know every sheep personally, etc. etc.

P. 85. I do not understand about the infant's sole; since all men become bipeds, all must exert pressure on sole. There is no disuse.

P. 88. Has not "muscardine" been substituted for "pebrine"? I have always considered this a very striking case. Here is apparent inheritance of a diseased state through the mother only, quite inexplicable till Pasteur discovered the rationale.

P. 155. Have you considered that State Socialism (for which I have little enough love) may be a product of Natural Selection? The societies of Bees and Ants exhibit socialism *in excelsis*.

The unlucky substitution of "survival of fittest" for "natural selection" has done much harm in consequence of the ambiguity of "fittest"—which many take to mean "best" or "highest"—whereas natural selection may work towards degradation: *vide epizoa*.

You do not refer to the male mamma—which becomes functional once in many million cases, see the curious records of Gynæcomasty. Here practical disuse in the male ever since the origin of the mammalia has not abolished the mamma or destroyed its functional potentiality in extremely rare cases.

I absolutely disbelieve in use-inheritance as the evidence stands. Spencer is bound to it *a priori*—his psychology goes to pieces without it.

Now as to the letter. I am no pessimist—but also no optimist. The world might be much worse, and it might be much better. Of moral purpose I see no trace in Nature. That is an article of exclusively human manufacture—and very much to our credit.

If you will accept the results of the experience of an old man who has had a very chequered existence—and has nothing to hope for except a few years of quiet down-hill—there is nothing of permanent value (putting aside a few human affections), nothing that satisfies quiet reflection—except the sense of having worked according to one's capacity and light, to make things clear and get rid of cant and shams of all sorts. That was the lesson I learned from Carlyle's books when I was a boy, and it has stuck by me all my life.

Therefore, my advice to you is go ahead. You may make more of failing to get money, and of succeeding in getting abuse—until such time in your life as (if you are teachable) you have ceased to care much about either. The job you propose to undertake is a big one, and will tax all your energies and all your patience.

But, if it were my case, I should take my chance of failing in a worthy task rather than of succeeding in lower things.

And if at any time I can be of use to you (even to the answering of letters) let me know. But in truth I am getting rusty in science—from disuse.—Ever yours very faithfully,   T. H. HUXLEY.

P.S.—Yes—Mr. Gladstone has dug up the hatchet. We shall see who gets the scalps.

By the way, you have not referred to plants, which are a stronghold for you. What is the good of use-inheritance, say, in orchids ?

The interests which had formerly been divided between biology and other branches of science and philosophy, were diverted from the one channel only to run stronger in the rest. Stagnation was the one thing impossible to him ; his rest was mental activity without excessive physical fatigue ; and he felt he still had a useful purpose to serve, as a friend put it, in patrolling his beat with a vigilant eye to the loose characters of thought. Thus he writes on September 29 to Sir J. Hooker :—

I wish quietude of mind were possible to me. But without something to do that amuses me and does not involve too much labour, I become quite unendurable—to myself and everybody else.

Providence has, I believe, specially devolved on Gladstone, Gore, and Co. the function of keeping "'ome 'appy" for me.

I really can't give up tormenting *ces drôles.*

However, I have been toiling at a tremendously scientific article about the "Aryan question" absolutely devoid of blasphemy.

This article appeared in the November number of the *Nineteenth Century* (*Coll. Essays*, vii. 271) and treats the question from a biological point of view, with the warning to readers that it is essentially a speculation based upon facts, but not assuredly proved. It starts from the racial characteristics of skull and stature, not from simply philological considerations, and arrives at a form of the "Sarmatian" theory of Aryan origins. And for fear lest he should be supposed to take sides in the question of race and language, or race and civilisation, he remarks :—

The combination of swarthiness with stature above the average and a long skull, confer upon me the serene impartiality of a mongrel.

THE GRAND HOTEL, EASTBOURNE,
Aug. 12, 1890.

MY DEAR EVANS—I have read your address returned herewith with a great deal of interest, as I happen to have been amusing myself lately with reviewing the "Aryan" question according to the new lights (or darknesses).

I have only two or three remarks to offer on the places I have marked A and B.

As to A, I would not state the case so strongly against the probabilities of finding pliocene man. A pliocene *Homo* skeleton might analogically be expected to differ no more from that of modern men than the Œningen *Canis* from modern *Canes*, or pliocene horses from modern horses. If so, he would most undoubtedly be a man— genus *Homo*—even if you made him a distinct species. For my part I should by no means be astonished to find the genus *Homo* represented in the Miocene, say the Neanderthal man with rather smaller brain capacity, longer arms and more movable great toe, but at most specifically different.

As to B, I rather think there were people who fought the fallacy of language being a test of race before Broca— among them thy servant—who got into considerable hot water on that subject for a lecture on the forefathers and forerunners of the English people, delivered in 1870. Taylor says that Cuno was the first to insist upon the proposition that race is not co-extensive with language in 1871. That is all stuff. The same thesis had been maintained before I took it up, but I cannot remember by whom.[1]

Won't you refer to the Blackmore Museum? I was very much struck with it when at Salisbury the other day.

Hope they gave you a better lunch at Gloucester than we did here. We'll treat you better next time in our own den. With the wife's kindest regards—Ever yours very faithfully,   T. H. HUXLEY.

The remark in a preceding letter about "Gladstone, Gore, and Co." turned out to be prophetic as well as retrospective. Mr. Gladstone published this autumn in *Good Words* his "Impregnable Rock of Holy Scripture," containing an attack upon Huxley's position as taken up in their previous controversy of 1889.

The debate now turned upon the story of the Gadarene swine. The question at issue was not, at first sight, one of vital importance, and one critic at least remarked that at their age Mr. Gladstone and Professor Huxley might be better occupied than in fighting over the Gadarene pigs :—

If these too famous swine were the only parties to the suit, I for my part (writes Huxley, *Coll. Essays*, v. 414) should fully admit the justice of the rebuke. But the real issue (he contends) is whether the men of the nineteenth century are to adopt the demonology of the men of the first century, as divinely revealed truth, or to reject it as degrading falsity.

---

[1] Cp. letter to Max Müller of June 15, 1865, vol. i. p. 380.

A lively encounter followed :—

The G.O.M. is not murdered (he writes on November 20), only "fillipped with a three-man beetle," as the fat knight has it.

This refers to the forthcoming article in the December *Nineteenth Century*, "The Keepers of the Herd of Swine," which was followed in March 1891 by "Mr. Gladstone's Controversial Methods" (see *Coll. Essays*, v. 366 *sqq.*), the rejoinder to Mr. Gladstone's reply in February.

The scope of this controversy was enlarged by the intervention in the January *Nineteenth Century* of the Duke of Argyll, to whom he devoted the concluding paragraphs of his March article. But it was scarcely well under way when another, accompanied by much greater effusion of ink and passion, sprang up in the columns of the *Times*. His share in it, published in 1891 as a pamphlet under the title of "Social Diseases and Worse Remedies," is to be found in *Coll. Essays*, ix. 237.

I have a new row on hand *in re* Salvation Army ! (he writes on December 2). It's all Mrs. ——'s fault; she offered the money.

In fact, a lady who was preparing to subscribe £1000 to "General" Booth's "Darkest England" scheme, begged Huxley first to give her his opinion of the scheme and the likelihood of its being properly carried out. A careful examination of "Darkest

England" and other authorities on the subject, convinced him that it was most unwise to create an organisation whose absolute obedience to an irresponsible leader might some day become a serious danger to the State; that the reforms proposed were already being undertaken by other bodies, which would be crippled if this scheme were floated; and that the financial arrangements of the Army were not such as provide guarantees for the proper administration of the funds subscribed :—

And if the thing goes on much longer, if Booth establishes his Bank, you will have a crash some of these fine days, comparable only to Law's Mississippi business, but unfortunately ruining only the poor.

On the same day he writes to his eldest son :—

HODESLEA, EASTBOURNE,
Dec. 8, 1890.

Attacking the Salvation Army may look like the advance of a forlorn hope, but this old dog has never yet let go after fixing his teeth into anything or anybody, and he is not going to begin now. And it is only a question of holding on. Look at Plumptre's letter exposing the Bank swindle.

The *Times*, too, is behaving like a brick. This world is not a very lovely place, but down at the bottom, as old Carlyle preached, veracity does really lie, and will show itself if people won't be impatient.

No sooner had he begun to express these opinions in the columns of the *Times* than additional information of all kinds poured in upon him, especially from within the Army, much of it private for fear of injury to the writers if it were discovered that they had written to expose abuses; indeed in one case the writer had thought better of even appending his signature to his letter, and had cut off his name from the foot of it, alleging that correspondence was not inviolable. So far were these persons from feeling hostility to the organisation to which they belonged, that one at least hailed the Professor as the divinely-appointed redeemer of the Army, whose criticism was to bring it back to its pristine purity.

To his Elder Son

HODESLEA, EASTBOURNE,
Jan. 8, 1891.

DEAR LENS—It is very jolly to think of J. and you paying us a visit. It is proper, also, the eldest son should hansel the house.

Is the Mr. Sidgwick who took up the cudgels for me so gallantly in the *St. James'* one of your Sidgwicks? If so, I wish you would thank him on my account. (The letter was capital.) [1] Generally people like me to pull

---

[1] Mr. William C. Sidgwick had written (January 4) an indignant letter to protest against the heading of an article in the *Speaker*, " Professor Huxley as Titus Oates." " To this monster of iniquity the *Speaker* compares an honourable English gentleman, because he has ventured to dissuade his countrymen from giving money to Mr. William Booth. . . . Mr. Huxley's views on theology may be wrong, but nobody doubts that he honestly holds them; they do not bring Mr. Huxley wealth and honours, nor do they cause the murder of the innocent. To insinuate a resemblance which you dare not state openly is an outrage on common decency. . . ."

the chestnuts out of the fire for them, but don't care to take any share in the burning of the fingers.

But the Boothites are hard hit, and may be allowed to cry out.

I begin to think that they must be right in saying that the Devil is at work to destroy them. No other theory sufficiently accounts for the way they play into my hands. Poor Clibborn-Booth has a long—columns long—letter in the *Times* to-day, in which, all unbeknownst to himself, he proves my case.

I do believe it is a veritable case of the herd of swine, and I shall have to admit the probability of that miracle.

Love to J. and Co. from us all.—Ever your affectionate

PATER.

HODESLEA, EASTBOURNE,
Jan. 11, 1891.

MY DEAR MR. CLODD—I am very much obliged to you for the number of the *St. James's Gazette*, which I had not seen. The leading article expresses exactly the same conclusions as those at which I had myself arrived from the study of the deed of 1878. But of course I was not going to entangle myself in a legal discussion. However, I have reason to know that the question will be dealt with by a highly qualified legal expert before long. The more I see of the operations of headquarters the worse they look. I get some of my most valuable information and heartiest encouragement from officers of the Salvation Army ; and I knew, in this way, of Smith's resignation a couple of days before it was announced ! But the poor fellows are so afraid of spies and consequent persecution, that some implore me not to notice their letters, and all pledge me to secrecy. So that I am Vice-Fontanelle with my hand full of truth, while I can only open my little finger.

It is a case of one down and t'other come on, just now. "——" will get his deserts in due time. But, oh dear, what a waste of time for a man who has not much to look to. No; "waste" is the wrong word; it's useful, but I wish that somebody else would do it and leave me to my books.

My wife desires her kind regards. I am happy to say she is now remarkably well. If you are this way, pray look in at our Hermitage.—Yours very faithfully,

T. H. HUXLEY.

HODESLEA, EASTBOURNE,
Jan. 30, 1891.

MY DEAR HOOKER—I trust I have done with Booth and Co. at last. What an ass a man is to try to prevent his fellow-creatures from being humbugged ! Surely I am old enough to know better. I have not been so well abused for an age. It's quite like old times.

And now I have to settle accounts with the duke and the G.O.M. I wonder when the wicked will let me be at peace.—Ever yours affectionately,   T. H. HUXLEY.

Other letters touch upon the politics of the hour, especially upon the sudden and dramatic fall of Parnell. He could not but admire the power and determination of the man, and his political methods, an admiration rashly interpreted by some journalist as admiration of the objects to which these political methods were applied. (See ii. 441.)

GRAND HOTEL, EASTBOURNE,
Nov. 26, 1890.

MY DEAR LECKY—Very. many thanks for your two volumes, which I rejoice to have, especially as a present from you. I was only waiting until we were settled in our new house—as I hope we shall be this time next week—to add them to the set which already adorn my shelves, and I promise myself soon to enjoy the reading of them.

The Unionist cause is looking up. What a strange thing it is that the Irish malcontents are always sold, one way or the other, by their leaders.

I wonder if the G.O.M. ever swears! Pity if he can't have that relief just now.

With our united kind regards to Mrs. Lecky and your-self—Ever yours very faithfully,   T. H. HUXLEY.

GRAND HOTEL, EASTBOURNE,
Nov. 29, 1890.

MY DEAR HOOKER—I have filled up and sent your and my copies of entry for Athenæum.

Carpenter has written the best popular statement I know of, of the results of criticism, in a little book called *The First Three Gospels,* which is well worth reading. [See p. 166.]

I have promised to go to R.S. dinner and propose Stokes' health on Monday, but if the weather holds out as Arctic as it is now, I shall not dare to venture. The driving east wind, blowing the snow before it here, has been awful; for ten years they have had nothing like it. I am glad to say that my little house turns out to be warm. We go in next Wednesday, and I fear I cannot be in town on Thursday even if the weather permits.

I have had pleurisy that was dangerous and not painful, then p. that was painful and not dangerous; there is only one further combination, and I don't want that.

Politics now are immensely interesting. There must be a depth of blackguardism in me, for I cannot help admiring Parnell. I prophesy that it is Gladstone who will retire for a while, and then come back to Parnell's heel like a whipped hound. His letter was carefully full of loopholes.—Ever yours affectionately,

T. H. HUXLEY.

HODESLEA, EASTBOURNE,
Dec. 2, 1890.

MY DEAR HOOKER— . . . The question of questions
now is whether the Unionists will have the sense to carry
a measure settling the land question at once.   If they do
that, I do not believe it will be in the power of man to
stir them further.   And my belief is that Parnell will be
quite content with that solution.   He does not want to
be made a nonentity by Davitt or the Irish Americans.

But what ingrained liars they all are !   That is the
bottom of all Irish trouble.   Fancy Healy and Sexton
going to Dublin to swear eternal fidelity to their leader,
and now openly declaring that they only did so because
they believed he would resign.—Ever yours affectionately,

T. H. HUXLEY.

HODESLEA, *Jan.* 10, 1891.

MY DEAR FOSTER—I am trying to bring the Booth
business to an end so far as I am concerned, but it's like
getting a wolf by the ears ; you can't let him go exactly
when you like.

But the result is quite worth the trouble   Booth,
Stead, Tillett, Manning and Co. have their little game
spoilt for the present.

You cannot imagine the quantity of letters I get from
the Salvation Army subordinates, thanking me and telling
me all sorts of stories in strict confidence.   The poor devils
are frightened out of their lives by headquarter spies.
Some beg me not to reply, as their letters are opened.

I knew that saints were not bad hands at lying before ;
but these Booth people beat Banagher.

Then there is —— awaits skinning, and I believe the
G.O.M. is to be upon me !   Oh for a quiet life.—Ever
yours faithfully,   T. H. HUXLEY.

But by February 17 the Booth business was over,

the final rejoinder to Mr. Gladstone sent to press;
and he writes to Sir J. Hooker :—

Please the pigs, I have now done with them—wiped
my mouth, and am going to be good—till next time.

But in truth I am as sick of controversy as a confec-
tioner's boy of tarts.

I rather think I shall set up as a political prophet.
Gladstone and all the rest are coming to heel to their
master.

Years ago one of the present leaders of the anti-
Parnellites said to me : " Gladstone is always in the hands
of somebody stronger than himself ; formerly it was Bright,
now it is Parnell."

CHAPTER VIII

1890–1891

THE new house at Eastbourne has been several times referred to. As usually happens, the move was considerably delayed by the slowness of the workmen; it did not actually take place till the beginning of December.

He writes to his daughter, Mrs. Roller, who also had just moved into a new house :—

You have all my sympathies on the buy, buy question. I never knew before that when you go into a new house money runs out at the heels of your boots. On former occasions, I have been too busy to observe the fact. But I am convinced now that it is a law of nature.

The origin of the name given to the house appears from the following letter :—

GRAND HOTEL, EASTBOURNE,
Oct. 15, 1890.

MY DEAR FOSTER—Best thanks for the third part of the " Physiology," which I found when I ran up to town for a day or two last week. What a grind that book must be.

How's a' wi' you ?　Let me have a line.

We ought to have been in our house a month ago, but fitters, paperers, and polishers are like bugs or cockroaches, you may easily get 'em in, but getting 'em out is the deuce.　However, I hope to clear them out by the end of this week, and get in by the end of next week.

One is obliged to have names for houses here.　Mine will be " Hodeslea," which is as near as I can go to " Hodesleia," the poetical original shape of my very ugly name.

There was a noble scion of the house of Huxley of Huxley who, having burgled and done other wrong things (temp. Henry IV.), asked for benefit of clergy.　I expect they gave it him, not in the way he wanted, but in the way they would like to " benefit " a later member of the family.

[Rough sketch of one priest hauling the rope taut over the gallows, while another holds a crucifix before the suspended criminal.]

Between this gentleman and my grandfather there is unfortunately a complete blank, but I have none the less faith in him as my ancestor.

My wife, I am sorry to say, is in town—superintending packing up—no stopping her.　I have been very uneasy about her at times, and shall be glad when we are quietly settled down.　With kindest regards to Mrs. Foster—Ever yours,　　　　　　　T. H. HUXLEY.

His own principal task was in getting his library ready for the move.

Most of my time (he writes on November 16) for the last fortnight has been spent in arranging books and tearing up papers till my back aches and my fingers are sore.

However, he did not take all his books with him. There was a quantity of biological works of all sorts which had accumulated in his library and which he was not likely to use again; these he offered as a parting gift to the Royal College of Science. On December 8, the Registrar conveys to him the thanks of the Council for "the valuable library of biological works," and further informs him that it was resolved—

That the library shall be kept in the room formerly occupied by the Dean, which shall be called "The Huxley Laboratory for Biological Research," and be devoted to the prosecution of original researches in Biological Science, with which the name of Professor Huxley is inseparably associated.

Huxley replied as follows :—

DEAR REGISTRAR—I beg you convey my hearty thanks to the Council for the great kindness of the minute and resolution which you have sent me. My mind has never been greatly set on posthumous fame; but there is no way of keeping memory green which I should like so well as that which they have adopted towards me.

It has been my fate to receive a good deal more vilipending than (I hope) I deserve. If my colleagues, with whom I have worked so long, put too high a value upon my services, perhaps the result may be not far off justice.—Yours very faithfully,   T. H. HUXLEY.

In addition to the directly controversial articles in the early part of the year, two other articles on controversial subjects belong to 1891. "Hasisadra's Adventure," published in the *Nineteenth Century* for June, completed his long-contemplated examination of the Flood myth. In this he first discussed the Babylonian form of the legend recorded upon the clay

tablets of Assurbanipal—a simpler and less ex-
aggerated form as befits an earlier version, and in its
physical details keeping much nearer to the bounds
of probability.

The greater part of the article, however, is
devoted to a wider question—How far does geological
and geographical evidence bear witness to the consequences which must have ensued from a universal
flood, or even from one limited to the countries of
Mesopotamia? And he comes to the conclusion that
these very countries have been singularly free from
any great changes of the kind for long geological
periods.

The sarcastic references in this article to those
singular reasoners who take the possibility of an
occurrence to be the same as scientific testimony to
the fact of its occurrence, lead up, more or less, to
the subject of an essay, "Possibilities and Impossibilities," which appeared in the *Agnostic Annual*
for 1892, actually published in October 1891, and to
be found in *Collected Essays*, v. 192.

This was a restatement of the fundamental
principles of the agnostic position, arising out of the
controversies of the last two years upon the demonology of the New Testament. The miraculous is not
to be denied as impossible ; as Hume said, " Whatever is intelligible and can be distinctly conceived
implies no contradiction, and can never be proved
false by any demonstrative argument or abstract
reasoning *a priori*," and these combinations of phen-

omena are perfectly conceivable. Moreover, in the progress of knowledge, the miracles of to day may be the science of to-morrow. Improbable they are, certainly, by all experience, and therefore they require specially strong evidence. But this is precisely what they lack ; the evidence for them, when examined, turns out to be of doubtful value.

I am anxious (he says) to bring about a clear understanding of the difference between " impossibilities " and " improbabilities," because mistakes on this point lay us open to the attacks of ecclesiastical apologists of the type of the late Cardinal Newman. . . .

When it is rightly stated, the Agnostic view of " miracles " is, in my judgment, unassailable. We are *not* justified in the *a priori* assertion that the order of nature, as experience has revealed it to us, cannot change. In arguing about the miraculous, the assumption is illegitimate, because it involves the whole point in dispute. Furthermore, it is an assumption which takes us beyond the range of our faculties. Obviously, no amount of past experience can warrant us in anything more than a correspondingly strong expectation for the present and future. We find, practically, that expectations, based upon careful observations of past events, are, as a rule, trustworthy. We should be foolish indeed not to follow the only guide we have through life. But, for all that, our highest and surest generalisations remain on the level of justifiable expectations ; that is, very high probabilities. For my part, I am unable to conceive of an intelligence shaped on the model of that of men, however superior it might be, which could be any better off than our own in this respect ; that is, which could possess logically justifiable grounds for certainty about the constancy of the order of things, and therefore be in a position to declare that such and such events are impossible. Some of the old mythologies recognised this clearly enough. Beyond and above Zeus and Odin, there lay the unknown and inscrutable Fate which, one day or other, would crumple up them and the world they ruled to give place to a new order of things.

I sincerely hope that I shall not be accused of Pyrrhonism, or of any desire to weaken the foundations of rational certainty. I have merely desired to point out that rational certainty is one thing, and talk about "impossibilities," or "violation of natural laws," another. Rational certainty rests upon two grounds; the one that the evidence in favour of a given statement is as good as it can be; the other, that such evidence is plainly insufficient. In the former case, the statement is to be taken as true, in the latter as untrue; until something arises to modify the verdict, which, however properly reached, may always be more or less wrong, the best information being never complete, and the best reasoning being liable to fallacy.

To quarrel with the uncertainty that besets us in intellectual affairs would be about as reasonable as to object to live one's life, with due thought for the morrow, because no man can be sure he will be alive an hour hence. Such are the conditions imposed upon us by nature, and we have to make the best of them. And I think that the greatest mistake those of us who are interested in the progress of free thought can make is to overlook these limitations, and to deck ourselves with the dogmatic feathers which are the traditional adornment of our opponents. Let us be content with rational certainty, leaving irrational certainties to those who like to muddle their minds with them.

As for the difficulty of believing miracles in themselves, he gives in this paper several examples of a favourite saying of his, that Science offers us much greater marvels than the miracles of theology; only the evidence for them is very different.

The following letter was written in acknowledgment of a paper by the Rev. E. McClure, which endeavoured to place the belief in an individual permanence upon the grounds that we know of no leakage anywhere in nature; that matter is not a source, but a transmitter of energy; and that the brain, so far from originating thought, is a mere machine responsive to something external to itself, a revealer of something which it does not produce, like a musical instrument. This "something" is the universal of thought, which is identified with the general λόγος of the fourth gospel. Moral perfection consists in assimilation to this; sin is the falling short of perfect revealing of the eternal λόγος.

Huxley's reply interested his correspondent not only for the brief opinion on the philosophic question, but for the personal touch in the explanation of the motives which had guided his life-work, and his "kind feeling towards such of the clergy as endeavoured to seek honestly for a natural basis to their faith."

HODESLEA, EASTBOURNE,
March 17, 1891.

DEAR MR. McCLURE—I am very much obliged for your letter, which belongs to a different category from most of those which I receive from your side of the hedge that, unfortunately, separates thinking men.

So far as I know myself, after making due deduction for the ambition of youth and a fiery temper, which ought to (but unfortunately does not) get cooler with age, my sole motive is to get at the truth in all things.

I do not care one straw about fame, present or posthumous, and I loathe notoriety, but I do care to have that desire manifest and recognised.

Your paper deals with a problem which has profoundly interested me for years, but which I take to be insoluble. It would need a book for full discussion. But I offer a remark only on two points.

The doctrine of the conservation of energy tells neither one way nor the other. Energy is the cause of movement of body, *i.e.* things having mass. States of consciousness have no mass, even if they can be conceded to be movable. Therefore even if they are caused by molecular movements, they would not in any way affect the store of energy.

Physical causation need not be the only kind of causation, and when Cabanis said that thought was a function of the brain, in the same way as bile secretion is a *function* of the liver, he blundered philosophically. Bile is a product of the transformation of material energy. But in the mathematical sense of the word "function," thought may be a function of the brain. That is to say, it may arise only when certain physical particles take on a certain order.

By way of a coarse analogy, consider a parallel-sided piece of glass through which light passes. It forms no picture. Shape it so as to be bi-convex, and a picture appears in its focus.

Is not the formation of the picture a "function" of the piece of glass thus shaped?

So, from your own point of view, suppose a mind-stuff —λόγος—a noumenal cosmic light such as is shadowed in the fourth gospel. The brain of a dog will convert it into one set of phenomenal pictures, and the brain of a man into another. But in both cases the result is the consequence of the way in which the respective brains perform their "functions."

Yet one point.

The actions we call sinful are as much the consequence of the order of nature as those we call virtuous. They are part and parcel of the struggle for existence through which all living things have passed, and they have become sins because man alone seeks a higher life in voluntary association.

Therefore the instrument has never been marred; on the contrary, we are trying to get music out of harps, sacbuts, and psalteries, which never were in tune and seemingly never will be.—Ever yours very faithfully,

T. H. HUXLEY.

Few years passed without some utterance from Huxley on the subject of education, especially scientific education. This year we have a letter to Professor Ray Lankester touching the science teaching at Oxford.

HODESLEA, EASTBOURNE,
Jan. 28, 1891.

DEAR LANKESTER—I met Foster at the Athenæum when I was in town last week, and we had some talk about your " very gentle " stirring of the Oxford pudding. I asked him to let you know when occasion offered, that (as I had already said to Burdon Sanderson) I drew a clear line apud biology between the medical student and the science student.

With respect to the former, I consider it ought to be kept within strict limits, and made simply a Vorschule to human anatomy and physiology.

On the other hand, the man who is going out in natural science ought to have a much larger dose, especially in the direction of morphology. However, from what I understood from Foster, there seems a doubt about the " going out " in Natural Science, so I had better confine

myself to the medicos. Their burden is already so heavy
that I do not want to see it increased by a needless
weight even of elementary biology.

Very many thanks for the "Zoological articles" just
arrived.— Ever yours very faithfully,

T. H. HUXLEY.

Don't write to the *Times* about anything ; look at the
trouble that comes upon a harmless man for two months,
in consequence.

The following letter, which I quote from the
Yorkshire Herald of April 11, 1891, was written in
answer to some inquiries from Mr. J. Harrison, who
read a paper on Technical Education as applied to
Agriculture, before the Easingwold Agricultural
Club :—

I am afraid that my opinion upon the subject of your
inquiry is worth very little—my ignorance of practical
agriculture being profound.   However, there are some
general principles which apply to all technical training ;
the first of these, I think, is that practice is to be learned
only by practice.   The farmer must be made by and
through farm work.   I believe I might be able to give
you a fair account of a bean plant and of the manner and
condition of its growth, but if I were to try to raise a
crop of beans, your club would probably laugh consumedly
at the result.   Nevertheless, I believe that you practical
people would be all the better for the scientific knowledge
which does not enable me to grow beans.   It would keep
you from attempting hopeless experiments, and would
enable you to take advantage of the innumerable hints
which Dame Nature gives to people who live in direct
contact with things.   And this leads me to the second
general principle which I think applies to all technical
teaching for school-boys and school-girls, and that is, that

they should be led from the observation of the commonest facts to general scientific truths.   If I were called upon to frame a course of elementary instruction preparatory to agriculture, I am not sure that I should attempt chemistry, or botany, or physiology or geology, as such.   It is a method fraught with the danger of spending too much time and attention on abstraction and theories, on words and notions instead of things.   The history of a bean, of a grain of wheat, of a turnip, of a sheep, of a pig, or of a cow properly treated—with the introduction of the elements of chemistry, physiology, and so on as they come in—would give all the elementary science which is needed for the comprehension of the processes of agriculture in a form easily assimilated by the youthful mind, which loathes everything in the shape of long words and abstract notions, and small blame to it.   I am afraid I shall not have helped you very much, but I believe that my suggestions, rough as they are, are in the right direction.

The remaining letters of the year are of miscellaneous interest.   They show him happily established in his retreat at Eastbourne in very fair health, on his guard against any further repetition of his "jubilee honour" in the shape of his old enemy pleurisy; unable to escape the more insidious attacks of influenza, but well enough on the whole to be in constant good spirits.

HODESLEA, EASTBOURNE,
Jan. 13, 1891.

MY DEAR SKELTON—Many thanks to you for reminding me that there are such things as "Summer Isles" in the universe.   The memory of them has been pretty well blotted out here for the last seven weeks.   You see some people can retire to "Hermitages" as well as other people;

and though even **Argyll** *cum* Gladstone powers of self-deception could not persuade me that the view from my window is as good as that from yours, yet I do see a fine wavy chalk down with "cwms" and soft turfy ridges, over which an old fellow can stride as far as his legs are good to carry him.

The fact is, that I discovered that staying in London any longer meant for me a very short life, and by no means a merry one. So I got my son-in-law to build me a cottage here, where my wife and I may go down-hill quietly together, and "make our sowls" as the Irish say, solaced by an occasional visit from children and **grand**-children.

The deuce of it is, that however much the weary want to be at rest the wicked won't cease from troubling. Hence the occasional skirmishes and alarms which may lead my friends to misdoubt my absolute detachment from sublunary affairs. Perhaps peace dwells only among the fork-tailed Petrels !

I trust Mrs. Skelton and you are flourishing, and that trouble will keep far from the hospitable doors of Braid through the New Year.—Ever yours very faithfully,

T. H. HUXLEY.

No sooner had he settled down in his new country home, than a strange piece of good fortune, such as happens more often in a story-book than in real life, enabled him at one stroke to double his little estate, to keep off the unwelcome approach of the speculative builder, and to give himself scope for the newly-discovered delights of the garden. The sale of the house in Marlborough Place covered the greater part of the cost of Hodeslea ; but almost on the very day on which the sale was concluded, he became the possessor of another house at Worthing by the death of Mr. Anthony Rich, the well-known antiquarian. An old man, almost alone in the world, his admiration for the great work done recently in natural science had long since led him to devise his property to Darwin and Huxley, to the one his private fortune, to the other his house and its contents, notably a very interesting library.

As a matter of feeling, Huxley was greatly disinclined to part with this house, Chapel Croft, as soon as it had come into his hands. A year earlier, he might have made it his home; but now he had settled down at Eastbourne, and Chapel Croft, as it stood, was unlikely to find a tenant. Accordingly he sold it early in July, and with the proceeds bought the piece of land adjoining his house. Thus he writes to Sir J. Hooker :—

HODESLEA, EASTBOURNE,
May 17, 1891.

MY DEAR HOOKER—My estate is somewhat of a white elephant. There is about a couple of acres of ground well situated and half of it in the shape of a very pretty lawn and shrubbery, but unluckily, in building the house, dear old Rich thought of his own convenience and not mine (very wrong of him !), and I cannot conceive anybody but an old bachelor or old maid living in it. I do not believe anybody would take it as it stands. No doubt the site is valuable, and it would be well worth while to anybody with plenty of cash to spare to build on to the house and make it useful. But I neither have the cash, nor do I want the bother. However, Waller is going to look at the place for me and see what can be done. It seems hardly decent to sell it at once ; and moreover the value is likely to increase. I suppose at present it is worth £2000, but that is only a guess.

Apropos of naval portrait gallery, can you tell me if there is a portrait of old John Richardson anywhere extant ? I always look upon him as the founder of my fortunes, and I want to hang him up (just over your head) on my chimney breast. Voici ! [sketch showing the position of the pictures above the fireplace] :—

By your fruits ye shall judge them ! My cold was influenza, I have been in the most preposterously weak state ever since ; and at last my wife lost patience and called in the doctor, who is screwing me up with nux vomica.

Sound wind and limb otherwise.—Ever yours affectionately, T. H. HUXLEY.

And again on July 3 :—

I have just been offered £2800 for Anthony Rich's place and have accepted it. It is probably worth £3000, but if I were to have it on my hands and sell by auction I should get no more out of the transaction.

I am greatly inclined to put some of the money into a piece of land—a Naboth's vineyard—in front of my house and turn horticulturist. I find nailing up creepers a delightful occupation.

In the same letter he describes two meetings with old friends :—

Last Friday I ran down to Hindhead to see Tyndall. He was very much better than I hoped to find him, after such a long and serious illness, quite bright and " Tyndalloid " and not aged as I feared he would be. . . . The local doctor happened to be there during my visit and spoke very confidently of his speedy recovery. The leg is all right again, and he even talks of Switzerland, but I begged Mrs. Tyndall to persuade him to keep quiet and within reach of home and skilled medical attendance.

Saturday to Monday we were at Down, after six or seven years' interruption of our wonted visits. It was very pleasant if rather sad. Mrs. Darwin is wonderfully well—naturally aged—but quite bright and cheerful as usual. Old Parslow turned up on Sunday, just eighty, but still fairly hale. *Fuimus fuimus!*

[Parslow was the old butler who had been in Mr. Darwin's service for many years.]

To his Daughter, Mrs. Roller

HODESLEA, EASTBOURNE,
May 5, 1891.

You dear people must have entered into a conspiracy, as I had letters from all yesterday. I have never been so set up before, and begin to think that fathers (like port) must improve in quality with age. (No irreverent jokes about their getting crusty, Miss.)

Julian and Joyce taken together may perhaps give a faint idea of my perfections as a child. I have not only a distinct recollection of being noticed on the score of my good looks, but my mother used to remind me painfully of them in my later years, looking at me mournfully and saying, " And you were such a pretty boy ! "

Much as he would have liked to visit the Maloja again this year, the state of his wife's health forbade such a long journey. He writes just after his attack of influenza to Sir M. Foster, who had been suffering in the same way :—

HODESLEA, *May* 12, 1891.

MY DEAR FOSTER—I was very glad to hear from you. Pray don't get attempting to do anything before you are set up again.

I am in a ridiculous state of weakness, and bless my stars that I have nothing to do. I find it troublesome to do even that.

I wish ballooning had advanced so far as to take people to Maloja, for I do not think my wife ought to undertake such a journey, and yet I believe the high air would do us both more good than anything else. . . .

The University of London scheme appears to be coming to grief, as I never doubted it would.—Ever yours,

T. H. HUXLEY.

So instead of going abroad, he stayed in East-bourne till the end of August, receiving a short visit from his old friend Jowett, who, though sadly enfeebled by age, still persisted in travelling by himself, and a longer visit from his elder son and his family. But from September 11 to the 26th he and his wife made a trip through the west country, starting from Salisbury, which had so delighted him the year before, and proceeding by way of the Wye valley, which they had not visited since their honeymoon, to Llangollen. The first stage on the return journey was Chester, whence they made pious pilgrimage to the cradle of his name, Old Huxley Hall, some nine miles from Chester. Incorporated with a modern farm-house, and forming the present kitchen, are some solid stone walls, part of the old manor-house, now no longer belonging to any one of the name. From here they went to Coventry, where he had lived as a boy, and found the house which his father had occupied still standing.

A letter to an old pupil contains reflections upon the years of work to which he had devoted so much of his energies.

TO PROFESSOR T. JEFFERY PARKER, OTAGO

HODESLEA, EASTBOURNE,
Aug. 11, 1891.

MY DEAR PARKER—It is a long time since your letter reached me, but I was so unwise as to put off answering it until the book arrived and I had read it. The book did not reach me for a long time, and what with one thing and another I have but just finished it. I assure you I am very proud of having my name connected with such a thorough piece of work, no less than touched by the kindness of the dedication.

Looking back from the aged point of view, the life which cost so much wear and tear in the living seems to have effected very little, and it is cheering to be reminded that one has been of some use.

Some years of continued ill-health, involving constant travelling about in search of better conditions than London affords, and long periods of prostration, have driven me quite out of touch with science. And indeed except for a certain toughness of constitution I should have been driven out of touch with terrestrial things altogether.

It is almost indecent in a man at my time of life who has had two attacks of pleurisy, followed by a dilated heart, to be not only above ground but fairly vigorous again. However, I am obliged to mind my P's and Q's; avoid everything like hard work, and live in good air.

The last condition we have achieved by setting up a

house close to the downs here ; and I begin to think with
Candide that "cultivons notre jardin" comprises the
whole duty of man.

I was just out of the way of hearing anything about
the University College chair ; and indeed, beyond attending the Council of the school when necessary, and meetings of Trustees of the British Museum, I rarely go to
London.

I have had my innings, and it is now for the younger
generation to have theirs.—With best wishes, ever yours
very faithfully,   T. H. HUXLEY.

As for being no longer in touch with the world of
science, he says the same thing in a note to Sir M.
Foster, forwarding an inquiry after a scientific teacher
(August 1)

Please read the enclosed, and if you know of anybody
suitable please send his name to Mr. Thomas.

I have told him that I am out of the way of knowing,
and that you are physiologically omniscient, so don't belie
the character !

This year a number of Huxley's essays were
translated into French. *Nature* for July 23, 1891
(vol. xliv. p. 272), notes the publication of "Les
Sciences Naturelles et l'Education," with a short
preface by himself, dwelling upon the astonishing
advance which had been made in the recognition of
science as an instrument of education, but warning
the younger generation that the battle is only half
won, and bidding them beware of relaxing their
efforts before the place of science is entirely assured.
In the issue for December 31 (*Nature*, xlvi. 397), is a

notice of " La Place de l'Homme dans la Nature," a re-issue of a translation of more than twenty years before, together with three ethnological essays, newly translated by M. H. de Varigny, to whom the following letters are addressed.

To H. DE VARIGNY

May 17, 1891.

I am writing to my publishers to send you *Lay Sermons, Critiques, Science and Culture,* and *American Addresses,* pray accept them in expression of my thanks for the pains you are taking about the translation. *Man's Place in Nature* has been out of print for years, so I cannot supply it.

I am quite conscious that the condensed and idiomatic English into which I always try to put my thoughts must present many difficulties to a translator. But a friend of mine who is a much better French scholar than I am, and who looked over two or three of the essays, told me he thought you had been remarkably successful.

The fact is that I have a great love and respect for my native tongue, and take great pains to use it properly. Sometimes I write essays half-a-dozen times before I can get them into the proper shape ; and I believe I become more fastidious as I grow older.

November 25, 1891.

I am very glad you have found your task pleasant, for I am afraid it must have cost you a good deal of trouble to put my ideas into the excellent French dress with which you have provided them. It fits so well that I feel almost as if I might be a candidate for a seat among the immortal forty !

As to the new volume, you shall have the refusal of it

if you care to have it. But I have my doubts about its acceptability to a French public which I imagine knows little about Bibliolatry and the ways of Protestant clericalism, and cares less.

These essays represent a controversy which has been going on for five or six years about Genesis, the deluge, the miracle of the herd of swine, and the miraculous generally, between Gladstone, the ecclesiastical principal of King's College, various bishops, the writer of *Lux Mundi*, that spoilt Scotch minister the Duke of Argyll, and myself.

My object has been to stir up my countrymen to think about these things; and the only use of controversy is that it appeals to their love of fighting, and secures their attention.

I shall be very glad to have your book on *Experimental Evolution*. I insisted on the necessity of obtaining experimental proof of the possibility of obtaining virtually infertile breeds from a common stock in 1860 (in one of the essays you have translated). Mr. Tegetmeier made a number of experiments with pigeons some years ago, but could obtain not the least approximation to infertility.

From the first, I told Darwin this was the weak point of his case from the point of view of scientific logic. But, in this matter, we are just where we were thirty years ago, and I am very glad you are going to call attention to the subject.

Sending a copy of the translation soon after to Sir J. Hooker, he writes :—

HODESLEA, EASTBOURNE,
Jan. 11, 1892.

MY DEAR HOOKER—We have been in the middle of snow for the last four days. I shall not venture to London, and if you deserve the family title of the "judicious," I don't think you will either.

I send you by this post a volume of the French translation of a collection of my essays about Darwinism and Evolution, 1860-76, for which I have written a brief preface. I was really proud of myself when I discovered on re-reading them that I had nothing to alter.

What times those days were ! *Fuimus !*—Ever yours affectionately,   T. H. HUXLEY.

The same subject of experimental evolution re-appears in a letter to Professor Romanes of April 29. A project was on foot for founding an institution in which experiments bearing upon the Darwinian theory could be carried out. After congratulating Professor Romanes upon his recent election to the Athenæum Club, he proceeds :—

In a review of Darwin's *Origin* published in the *Westminster* for 1860 (*Lay Sermons*, pp. 323-24), you will see that I insisted on the logical incompleteness of the theory so long as it was not backed by experimental proof that the cause assumed was competent to produce all the effects required. (See also *Lectures to Working Men*, 1863, pp. 146, 147.) In fact, Darwin used to reproach me sometimes for my pertinacious insistence on the need of experimental verification.

But I hope you are going to choose some other title than "Institut transformiste," which implies that the Institute is pledged to a foregone conclusion, that it is a workshop devoted to the production of a particular kind of article. Moreover, I should say that as a matter of prudence, you had better keep clear of the word "experimental." Would not "Biological Observatory" serve the turn ? Of course it does not exclude experiment any more than "Astronomical Observatory" excludes spectrum analysis.

Please think over this. My objection to "Transformist" is very strong.

In August his youngest daughter wrote to him to find out the nature of various "objects of the seashore" which she had found on the beach in South Wales. His answers make one wish that there had been more questions.

HODESLEA, EASTBOURNE,
Aug. 14, 1891.

DEAREST BABS—1. "Ornary" or not "ornary" B is merely A turned upside down and viewed with the imperfect appreciation of the mere artistic eye!

2. Your little yellow things are, I expect, egg-cases of dog whelks. You will find a lot of small eggs inside them, one or two of which grow faster than the rest, and eat up their weaker brothers and sisters.

The dog whelk is common on the shores. If you look for something like this [sketch of a terrier coming out of a whelk shell], you will be sure to recognise it.

3. Starfish are *not* born in their proper shape and don't come from your whitish yellow lumps. The thing that comes out of a starfish egg is something like this [sketch], and swims about by its cilia. The starfish proper is formed inside, and it is carried on its back this-uns.

Finally starfish drops off carrying with it t'other one's stomach, so that the subsequent proceedings interest t'other one no more.

4. The ropy sand tubes that make a sort of banks and reefs are houses of worms, that they build up out of sand, shells, and slime. If you knock a lot to pieces you will find worms inside.

5. Now, how do I know what the rooks eat? But there are a lot of unconsidered trifles about, and if you

get a good telescope and watch, you will have a glimpse as they hover between sand and rooks' beaks.

It has been blowing more or less of a gale here from the west for weeks—usually cold, often foggy—so that it seems as if summer were going to be late, probably about November.

But we thrive fairly well. L. and J. and their chicks are here and seem to stand the inclemency of the weather pretty fairly. The children are very entertaining.

M—— has been a little complaining, but is as active as usual.

My love to Joyce, and tell her I am glad to hear she has not forgotten her astronomy.

In answer to your inquiry, Leonard says that Trevenen has twenty-five teeth. I have a sort of notion this can be hardly accurate, but never having been a mother can't presume to say.—Our best love to you all.—Ever your loving PATER.

HODESLEA, EASTBOURNE,
Aug. 26, 1891.

DEAREST BABS—'Pears to me your friend is a squid or pen-and-ink fish, *Loligo* among the learned. Probably *Loligo media* which I have taken in that region. They have ten tentacles with suckers round their heads, two much longer than the others. They are close to cuttle-fish, but have a thin horny shell inside them instead of the "cuttle-bone." If you can get one by itself in a tub of water, it is pretty to see how they blush all over and go pale again, owing to little colour-bags in the skin, which expand and contract. Doubtless they took you for a heron, under the circumstances [sketch of a wader].

With slight intervals it has been blowing a gale from the west here for some months, the memory of man indeed goeth not back to the calm. I have not been really warm more than two days this so-called summer. And everybody prophesied we should be roasted alive here in summer.

We are all flourishing, and send our best love to Jack and you. Tell Joyce the wallflowers have grown quite high in her garden.—Ever your loving   PATER.

Politics are not often touched upon in the letters of this period, but an extract from a letter of October 25, 1891, is of interest as giving his reason for supporting a Unionist Government, many of whose tendencies he was far from sympathising with :—

The extract from the *Guardian* is wonderful. The Gladstonian tee-to-tum cannot have many more revolutions to make. The only thing left for him now, is to turn Agnostic, declare Homer to be an old bloke of a ballad-monger, and agitate for the prohibition of the study of Greek in all universities. . . .

It is just because I do not want to see our children involved in civil war that I postpone all political considerations to keeping up a Unionist Government.

I may be quite wrong; but right or wrong, it is no question of party. "Rads delight not me nor Tories neither," as Hamlet does not say.

The following letter to Sir M. Foster shows how little Huxley was now able to do in the way of public business without being knocked up :—

HODESLEA, *Oct.* 20, 1891.

MY DEAR FOSTER—If I had known the nature of the proceedings at the College of Physicians yesterday, I should have braved the tedium of listening to a lecture I could not hear in order to see you decorated. Clark had made a point of my going to the dinner,[1] and, worse luck,

---

[1] *I.e.* at the College of Physicians.

I had to "say a few words" after it, with the result that I am entirely washed out to-day, and only able to send you the feeblest of congratulations.—Ever yours,

T. H. Huxley.

The same thing appears in the following to Sir W. H. Flower, which is also interesting for his opinion on the question of promotion by seniority :—

HODESLEA, EASTBOURNE,
Oct. 23, 1891.

My dear Flower — My "next worst thing" was promoting a weak man to a place of responsibility in lieu of a strong one, on the mere ground of seniority.

Caeteris paribus, or with even approximate equality of qualifications, no doubt seniority ought to count ; but it is mere ruin to any service to let it interfere with the promotion of men of marked superiority, especially in the case of offices which involve much responsibility.

I suppose as trustee I may requisition a copy of Woodward's Catalogue. I should like to look a little more carefully at it. . . . We are none the worse for our pleasant glimpse of the world (and his wife) at your house ; but I find that speechifying at public dinners is one of the luxuries that I must utterly deny myself. It will take me three weeks' quiet to get over my escapade. —Ever yours very faithfully, T. H. Huxley.

CHAPTER IX

1892

THE revival of part of the former controversy which he had had with Mr. Gladstone upon the story of creation, made a warlike beginning of an otherwise very peaceful year. Since the middle of December a great correspondence had been going on in the *Times*, consequent upon the famous manifesto of the thirty-eight Anglican clergy touching the question of inspiration and the infallibility of the Bible. Criticism, whether "higher" or otherwise, defended on the one side, was unsparingly denounced on the other. After about a month of this correspondence, Huxley's name was mentioned as one of these critics; whereupon he was attacked by one of the disputants for "misleading the public" by his assertion in the original controversy that while reptiles appear in the geological record before birds, Genesis affirms the contrary; the critic declaring that the word for "creeping things" (rehmes) created on the sixth day, does not refer to reptiles, which are covered by the

" moving creatures " (shehretz) used of the first appearance of animal life.

It is interesting to see how, in his reply, Huxley took care to keep the main points at issue separate from the subordinate and unimportant ones. His answer is broken up into four letters. The first (*Times*, January 26) rehearses the original issue between himself and Mr. Gladstone; wherein both sides agreed that the creation of the sixth day included reptiles, so that, formally at least, his position was secure, though there was also a broader ground of difference to be considered. Before proceeding further, he asks his critic whether he admits the existence of the contradiction involved, and if not, to state his reasons therefor. These reasons were again given on February 1 as the new interpretation of the two Hebrew words already referred to, an interpretation, by the way, which makes the same word stand both for "the vast and various population of the waters " and "for such land animals as mice, weasels, and lizards, great and small."

On February 3 appeared the second letter, in which, setting aside the particular form which his argument against Mr. Gladstone had taken, he described the broad differences between the teachings of Genesis and the teachings of evolution. He left the minor details as to the interpretation of the words in dispute, which did not really affect the main argument, to be dealt with in the next letter of February 4. It was a question with which he had

long been familiar, as twenty years before he had, at
Dr. Kalisch's request, gone over the proofs of his
Commentary on Leviticus.

The letter of February 3 is as follows :—

While desirous to waste neither your space nor my
own time upon mere misrepresentations of what I have
said elsewhere about the relations between modern science
and the so-called "Mosaic" cosmogony, it seems needful
that I should ask for the opportunity of stating the case
once more, as briefly and fairly as I can.

I conceive the first chapter of Genesis to teach—(1)
that the species of plants and animals owe their origin to
supernatural acts of creation ; (2) that these acts took
place at such times and in such a manner that all the
plants were created first, all the aquatic and aerial
animals (notably birds) next, and all terrestrial animals
last.   I am not aware that any Hebrew scholar denies
that these propositions agree with the natural sense of
the text.   Sixty years ago I was taught, as most people
were then taught, that they are guaranteed by Divine
authority.

On the other hand, in my judgment, natural science
teaches no less distinctly—(1) that the species of animals
and plants have originated by a process of natural evolution ; (2) that this process has taken place in such a
manner that the species of animals and plants, respectively,
have come into existence one after another throughout the
whole period since they began to exist on the earth ; that
the species of plants and animals known to us are, as a
whole, neither older nor younger the one than the other.

The same holds good of aquatic and aerial species, as
a whole, compared with terrestrial species ; but birds
appear in the geological record later than terrestrial
reptiles, and there is every reason to believe that they
were evolved from the latter.

Until it is shown that the first two propositions are not contained in the first chapter of Genesis, and that the second pair are not justified by the present condition of our knowledge, I must continue to maintain that natural science and the "Mosaic" account of the origin of animals and plants are in irreconcilable antagonism.

As I greatly desire that this broad issue should not be obscured by the discussion of minor points, I propose to defer what I may have to say about the great "shehretz" and "rehmes" question till to-morrow.

On February 11 he wrote once more, again taking certain broader aspects of the problem presented by the first chapter of Genesis. He expressed his belief, as he had expressed it in 1869, that theism is not logically antagonistic to evolution. If, he continues, the account in Genesis, as Philo of Alexandria held, is only a poem or allegory, where is the proof that any one non-natural interpretation is the right one? and he concludes by pointing out the difficulties in the way of those who, like the famous thirty-eight, assert the infallibility of the Bible as guaranteed by the infallibility of the Church.

Apart from letters and occasional controversy, he published this year only one magazine article and a single volume of collected essays, though he was busy preparing the Romanes Lecture for 1893, the more so because there was some chance that Mr. Gladstone would be unable to deliver the first of the lectures in 1892, and Huxley had promised to be ready to take his place if necessary.

The volume (called *Controverted Questions*) which

appeared in 1892, was a collection of the essays of the last few years, mainly controversial, or as he playfully called them, "endeavours to defend a cherished cause," dealing with agnosticism and the demonological and miraculous element in Christianity. That they were controversial in tone no one lamented more than himself; and as in the letter to M. de Varigny, of November 25, 1891, so here in the prologue he apologises for the fact.

This prologue,—of which he writes to a friend, "It cost me more time and pains than any equal number of pages I have ever written,"—was designed to indicate the main question, various aspects of which are dealt with by these seemingly disconnected essays.

The historical evolution of humanity (he writes), which is generally, and I venture to think not unreasonably, regarded as progress, has been, and is being, accompanied by a co-ordinate elimination of the supernatural from its originally large occupation of men's thought. The question—How far is this process to go? is, in my apprehension, the controverted question of our time.

This movement, marked by the claim for the freedom of private judgment, which first came to its fulness in the Renascence, is here sketched out, rising or sinking by turns under the pressure of social and political vicissitudes, from Wiclif's earliest proposal to reduce the Supernaturalism of Christianity within the limits sanctioned by the Scriptures, down to the manifesto in the previous year of the thirty-eight Anglican divines in defence of biblical infallibility, which practically ends in an appeal to the very principle they reject.

But he does not content himself with pointing out the destructive effects of criticism upon the evidence in favour of a " supernature "—" The present incarnation of the spirit of the Renascence," he writes, "differs from its predecessor in the eighteenth century, in that it builds up, as well as pulls down. That of which it has laid the foundation, of which it is already raising the superstructure, is the doctrine of evolution," a doctrine that " is no speculation, but a generalisation of certain facts, which may be observed by any one who will take the necessary trouble." And in a short dozen pages he sketches out that " common body of established truths " to which it is his confident belief that " all future philosophical and theological speculations will have to accommodate themselves."

There is no need to recapitulate these ; they may be read in *Science and Christian Tradition*, the fifth volume of the Collected Essays ; but it is worth noticing that in conclusion, after rejecting " a great many supernaturalistic theories and legends which have no better foundations than those of heathenism," he declares himself as far from wishing to " throw the Bible aside as so much waste paper " as he was at the establishment of the School Board in 1870. As English literature, as world-old history, as moral teaching, as the *Magna Charta* of the poor and of the oppressed, the most democratic book in the world, he could not spare it. "I do not say," he adds, "that even the highest biblical ideal is exclusive of others or needs no supplement. But I do believe that the human race is not yet, possibly may never be, in a position to dispense with it."

It was this volume that led to the writing of the magazine article referred to above. The republication in it of the "Agnosticism," originally written in reply to an article of Mr. Frederic Harrison's, induced the latter to disclaim in the *Fortnightly Review* the intimate connection assumed to exist between his views and the system of Positivism detailed by Comte, and at the same time to offer the olive branch to his former opponent. But while gratefully accepting the goodwill implied in the offer, Huxley still declared himself unable to "give his assent to a single doctrine which is the peculiar property of Positivism, old or new," nor to agree with Mr. Harrison when he wanted

to persuade us that agnosticism is only the Court of the Gentiles of the Positivist temple ; and that those who profess ignorance about the proper solution of certain speculative problems ought to call themselves Positivists of the Gate, if it happens that they also take a lively interest in social and political questions.

This essay, "An Apologetic Irenicon," contains more than one passage of personal interest, which are the more worth quoting here, as the essay has not been republished. It was to have been included in a tenth volume of collected Essays, along with a number of others which he projected, but never wrote.

Thus, begging the Positivists not to regard him as a rival or competitor in the business of instructing the human race, he says :—

I aspire to no such elevated and difficult situation. I declare myself not only undesirous of it, but deeply conscious of a constitutional unfitness for it. Age and hygienic necessities bind me to a somewhat anchoritic life in pure air, with abundant leisure to meditate upon the wisdom of Candide's sage aphorism, "Cultivons notre jardin"—especially if the term garden may be taken broadly and applied to the stony and weed-grown ground within my skull, as well as to a few perches of more promising chalk down outside it. In addition to these effectual bars to any of the ambitious pretensions ascribed to me, there is another : of all possible positions that of master of a school, or leader of a sect, or chief of a party, appears to me to be the most undesirable ; in fact, the average British matron cannot look upon followers with a more evil eye than I do. Such acquaintance with the history of thought as I possess, has taught me to regard schools, parties, and sects, as arrangements, the usual effect of which is to perpetuate all that is worst and feeblest in the master's, leader's, or founder's work ; or else, as in some cases, to upset it altogether ; as a sort of hydrants for extinguishing the fire of genius, and for stifling the flame of high aspirations, the kindling of which has been the chief, perhaps the only, merit of the protagonist of the movement. I have always been, am, and propose to remain a mere scholar. All that I have ever proposed to myself is to say, this and this have I learned ; thus and thus have I learned it : go thou and learn better ; but do not thrust on my shoulders the responsibility for your own laziness if you elect to take, on my authority, conclusions, the value of which you ought to have tested for yourself.

Again, replying to the reproach that all his public utterances had been of a negative character, that the great problems of human life had been entirely left out of his purview, he defends once more the work of the man who clears the ground for the builders to come after him :—

> There is endless backwoodsman's work yet to be done. If "those also serve who only stand and wait," still more do those who sweep and cleanse ; and if any man elect to give his strength to the weeder's and scavenger's occupation, I remain of the opinion that his service should be counted acceptable, and that no one has a right to ask more of him than faithful performance of the duties he has undertaken.   I venture to count it an improbable suggestion that any such person—a man, let us say, who has well-nigh reached his threescore years and ten, and has graduated in all the faculties of human relationships ; who has taken his share in all the deep joys and deeper anxieties which cling about them ; who has felt the burden of young lives entrusted to his care, and has stood alone with his dead before the abyss of the eternal—has never had a thought beyond negative criticism.   It seems to me incredible that such an one can have done his day's work, always with a light heart, with no sense of responsibility, no terror of that which may appear when the factitious veil of Isis—the thick web of fiction man has woven round nature—is stripped off.

Challenged to state his " mental bias, *pro* or *con*," with regard to such matters as Creation, Providence, etc., he reiterates his words written thirty-two years before :—

So far back as 1860 I wrote :—

"The doctrine of special creation owes its existence very largely to the supposed necessity of making science accord with the Hebrew cosmogony"; and that the hypothesis of special creation is, in my judgment, a "mere specious mask for our ignorance." Not content with negation, I said :—

"Harmonious order governing eternally continuous progress ; the web and woof of matter and force interweaving by slow degrees, without a broken thread, that veil which lies between us and the infinite ; that universe which alone we know, or can know ; such is the picture which science draws of the world."

. . Every reader of Goethe will know that the second is little more than a paraphrase of the well-known utterance of the "Zeitgeist" in *Faust*, which surely is something more than a mere negation of the clumsy anthropomorphism of special creation.

Follows a query about "Providence," my answer to which must depend upon what my questioner means by that substantive, whether alone, or qualified by the adjective "moral."

If the doctrine of a Providence is to be taken as the expression, in a way "to be understanded of the people," of the total exclusion of chance from a place even in the most insignificant corner of Nature, if it means the strong conviction that the cosmic process is rational, and the faith that, throughout all duration, unbroken order has reigned in the universe, I not only accept it, but I am disposed to think it the most important of all truths. As it is of more consequence for a citizen to know the law than to be personally acquainted with the features of those who will surely carry it into effect, so this very positive doctrine of Providence, in the sense defined, seems to me far more important than all the theorems of speculative theology.   If, further, the doctrine is held to imply that, in some indefinitely remote past aeon, the cosmic process was set going by some entity possessed of intelligence and foresight, similar to our own in kind, however superior in degree, if, consequently, it is held that every event, not merely in our planetary speck, but in untold millions of other worlds, was foreknown before these worlds were, scientific thought, so far as I know anything about it, has nothing to say against that hypothesis.   It is, in fact, an anthropomorphic rendering of the doctrine of evolution.

It may be so, but the evidence accessible to us is, to my mind, wholly insufficient to warrant either a positive or a negative conclusion.

He remarks in passing upon the entire exclusion of "special" providences by this conception of a universal "Providence."   As for "moral" providence :—

So far as mankind has acquired the conviction that the observance of certain rules of conduct is essential to the maintenance of social existence, it may be proper to say that " Providence," operating through men, has generated morality.   Within the limits of a fraction of a fraction of the living world, therefore, there is a "moral" providence.   Through this small plot of an infinitesimal fragment of the universe there runs a "stream of tendency towards righteousness."   But outside the very rudimentary germ of a garden of Eden, thus watered, I am unable to discover any "moral" purpose, or anything but a stream of purpose towards the consummation of the cosmic process, chiefly by means of the struggle for existence, which is no more righteous or unrighteous than the operation of any other mechanism.

This, of course, is the underlying principle of the

Romanes Lecture, upon which he was still at work. It is more specifically expressed in the succeeding paragraph :—

I hear much of the "ethics of evolution." I apprehend that, in the broadest sense of the term "evolution," there neither is, nor can be, any such thing. The notion that the doctrine of evolution can furnish a foundation for morals seems to me to be an illusion which has arisen from the unfortunate ambiguity of the term "fittest" in the formula, "survival of the fittest." We commonly use "fittest" in a good sense, with an understood connotation of "best"; and "best" we are apt to take in its ethical sense. But the "fittest" which survives in the struggle for existence may be, and often is, the ethically worst.

Another paragraph explains the sense in which he used to say that the Romanes Lecture was a very orthodox discourse on the text, "Satan, the Prince of this world" :—

It is the secret of the superiority of the best theological teachers to the majority of their opponents that they substantially recognise these realities of things, however strange the forms in which they clothe their conceptions. The doctrines of predestination, of original sin, of the innate depravity of man and the evil fate of the greater part of the race, of the primacy of Satan in this world, of the essential vileness of matter, of a malevolent Demiurgus subordinate to a benevolent Almighty, who has only lately revealed himself, faulty as they are, appear to me to be vastly nearer the truth than the "liberal" popular illusions that babies are all born good, and that the example of a corrupt society is responsible for their failure to remain so ; that it is given to everybody to reach the ethical ideal if he will only try ; that all partial evil

is universal good, and other optimistic figments, such as
that which represents " Providence " under the guise of a
paternal philanthropist, and bids us believe that everything will come right (according to our notions) at last.

As to " Immortality " again [he refers his critic to his
book on " Hume "]. I do not think I need return to
" subjective " immortality, but it may be well to add that
I am a very strong believer in the punishment of certain
kinds of actions, not only in the present, but in all the
future a man can have, be it long or short. Therefore in
hell, for I suppose that all men with a clear sense of right
and wrong (and I am not sure that any others deserve
such punishment) have now and then " descended into
hell " and stopped there quite long enough to know what
infinite punishment means. And if a genuine, not merely
subjective, immortality awaits us, I conceive that, without
some such change as that depicted in the fifteenth chapter
of the first Epistle to the Corinthians, immortality must
be eternal misery. The fate of Swift's Struldbrugs seems
to me not more horrible than that of a mind imprisoned
for ever within the *flammantia mœnia* of inextinguishable
memories.

Further, it may be well to remember that the highest
level of moral aspiration recorded in history was reached
by a few ancient Jews—Micah, Isaiah, and the rest—who
took no count whatever of what might or might not
happen to them after death. It is not obvious to me
why the same point should not by and by be reached by
the Gentiles.

He admits that the generality of mankind will
not be satisfied to be told that there are some topics
about which we know nothing now, and do not seem
likely ever to be able to know more ; and, consequently, that in the long-run the world will turn to
those who profess to have conclusions :—

And that is the pity of it. As in the past, so, I fear, through a very long future, the multitude will continue to turn to those who are ready to feed it with the viands its soul lusteth after ; who will offer mental peace where there is no peace, and lap it in the luxury of pleasant delusions.

To missionaries of the Neo-Positivist, as to those of other professed solutions of insoluble mysteries, whose souls are bound up in the success of their sectarian propaganda, no doubt, it must be very disheartening if the "world," for whose assent and approbation they sue, stops its ears and turns its back upon them. But what does it signify to any one who does not happen to be a missionary of any sect, philosophical or religious, and who, if he were, would have no sermon to preach except from the text with which Descartes, to go no further back, furnished us two centuries since ? I am very sorry if people will not listen to those who rehearse before them the best lessons they have been able to learn, but that is their business, not mine. Belief in majorities is not rooted in my breast, and if all the world were against me the fact might warn me to revise and criticise my opinions, but would not in itself supply a ghost of a reason for forsaking them. For myself I say deliberately, it is better to have a millstone tied round the neck and be thrown into the sea than to share the enterprises of those to whom the world has turned, and will turn, because they minister to its weaknesses and cover up the awful realities which it shudders to look at.

A letter to Mr. N. P. Clayton also discusses the basis of morality.

HODESLEA, EASTBOURNE, *Nov.* 5, 1892.

DEAR SIR—I well remember the interview to which you refer, and I should have replied to your letter sooner, but during the last few weeks I have been very busy.

Moral duty consists in the observance of those rules of conduct which contribute to the welfare of society, and by implication, of the individuals who compose it.

The end of society is peace and mutual protection, so that the individual may reach the fullest and highest life attainable by man. The rules of conduct by which this end is to be attained are discoverable—like the other so-called laws of Nature—by observation and experiment, and only in that way.

Some thousands of years of such experience have led to the generalisations, that stealing and murder, for example, are inconsistent with the ends of society. There is no more doubt that they are so than that unsupported stones tend to fall. The man who steals or murders, breaks his implied contract with society, and forfeits all protection. He becomes an outlaw, to be dealt with as any other feral creature. Criminal law indicates the ways which have proved most convenient for dealing with him.

All this would be true if men had no "moral sense" at all, just as there are rules of perspective which must be strictly observed by a draughtsman, and are quite independent of his having any artistic sense.

The moral sense is a very complex affair—dependent in part upon associations of pleasure and pain, approbation and disapprobation formed by education in early youth, but in part also on an innate sense of moral beauty and ugliness (how originated need not be discussed), which is possessed by some people in great strength, while some are totally devoid of it—just as some children draw, or are enchanted by music while mere infants, while others do not know "Cherry Ripe" from "Rule Britannia," nor can represent the form of the simplest thing to the end of their lives.

Now for this last sort of people there is no reason why they should discharge any moral duty, except from fear of punishment in all its grades, from mere disapprobation to hanging, and the duty of society is to see that they live under wholesome fear of such punishment short, sharp, and decisive.

For the people with a keen innate sense of moral beauty there is no need of any other motive. What they want is knowledge of the things they may do and must leave undone, if the welfare of society is to be attained. Good people so often forget this that some of them occasionally require hanging almost as much as the bad.

If you ask why the moral inner sense is to be (under due limitations) obeyed ; why the few who are steered by it move the mass in whom it is weak ? I can only reply by putting another question—Why do the few in whom the sense of beauty is strong—Shakespere, Raffaele, Beethoven, carry the less endowed multitude away ? But they do, and always will. People who overlook that fact attend neither to history nor to what goes on about them.

Benjamin Franklin was a shrewd, excellent, kindly man. I have a great respect for him. The force of genial common-sense respectability could no further go. George Fox was the very antipodes of all this, and yet one understands how he came to move the world of his day, and Franklin did not.

As to whether we can all fulfil the moral law, I should say hardly any of us. Some of us are utterly incapable of fulfilling its plainest dictates. As there are men born physically cripples, and intellectually idiots, so there are some who are moral cripples and idiots, and can be kept straight not even by punishment. For these people there is nothing but shutting up, or extirpation.—I am, yours faithfully,   T. H. HUXLEY.

The peaceful aspect of the "Irenicon" seems to have veiled to most readers the unbroken nature of his defence, and he writes to his son-in-law, the Hon.

John Collier, suggesting an alteration in the title of
the essay:—

HODESLEA, EASTBOURNE,
Nov. 8, 1892.

MY DEAR JACK—It is delightful to find a reader who
"twigs" every point as acutely as your brother has done.
I told somebody—was it you?—I rather wished the
printer would substitute *o* for *e* in Irenicon. So far as I
have seen any notices, the British critic (what a dull ass
he is) appears to have been seriously struck by my sweetness of temper.

I sent you the article yesterday, so you will judge for
yourself.—With love, ever yours affectionately,

T. H. HUXLEY.

You should see the place I am claiming for Art in the
University. I do believe something will grow out of my
plan, which has made all the dry bones rattle. It is
coming on for discussion in the Senate, and I shall be
coming to you to have my wounds dressed after the fight.
Don't know the day yet.

This allusion to the place of Art in the University
refers to the proposed reorganisation of the London
University.

Since the year 1887 the question of establishing a
Teaching University for London had become more
and more pressing. London contained many isolated
teaching bodies of various kinds—University College,
King's College, the Royal College of Science, the
Medical Schools, Bedford College, and so forth, while
the London University was only an examining body.
Clearly these scattered bodies needed organising; the
the educational forces of the metropolis were dis-

integrated; much teaching—and this was especially true of the medical schools—that could have been better done and better paid in a single institution, was split up among several, none of which, perhaps, could offer sufficient inducement to keep the best men permanently.

The most burning question was, whether these bodies should be united into a new university, with power to grant degrees of its own, or should combine with the existing University of London, so that the latter would become a teaching as well as an examining body.   And if so, there was the additional question as to the form which this combination should take—whether federation, for example, or absorption.

The whole question had been referred to a Royal Commission by the Government of Lord Salisbury. The results were seen in the charter for a Gresham University, embodying the former alternative, and in the introduction into Parliament of a Bill to carry this scheme into effect.   But this action had only been promoted by some of the bodies interested, and was strongly opposed by other bodies, as well as by many teachers who were interested in university reform.

Thus at the end of February, Huxley was invited, as a Governor of University College, to sign a protest against the provisions of the Charter for a Teaching University then before Parliament, especially in so far as it was proposed to establish a second examining body in London.   The signatories also begged the

Government to grant further inquiry before legislating on the subject.

The protest, which received over 100 signatures of weight, contributed something towards the rejection of the Bill in the House of Commons. It became possible to hope that there might be established in London a University which should be something more than a mere collection of teachers, having as their only bond of union the preparation of students for a common examination. It was proposed to form an association to assist in the promotion of a teaching university for the metropolis ; but the first draft of a scheme to reconcile the complication of interests and ideals involved led Huxley to express himself as follows :—

HODESLEA, EASTBOURNE,
March 27, 1892.

DEAR PROFESSOR WELDON [1]—I am sorry to have kept you waiting so long for an answer to your letter of the 17th : but your proposal required a good deal of consideration, and I have had a variety of distractions.

So long as I am a member of the Senate of the University of London, I do not think I can with propriety join any Association which proposes to meddle with it. Moreover, though I have a good deal of sympathy with the ends of the Association, I have my doubts about many propositions set forth in your draft.

I took part in the discussions preliminary to Lord

---

[1] Then at University College, London ; now Linacre Professor of Physiology at Oxford.

Justice Fry's scheme, and I was so convinced that that scheme would be wrecked amidst the complication of interests and ideals that claimed consideration, that I gave up attending to it. In fact, living so much out of the world now, and being sadly deaf, I am really unfit to intervene in business of this kind.

Worse still, I am conscious that my own ideal is, for the present at any rate, hopelessly impracticable. I should cut away medicine, law, and theology as technical specialities in charge of corporations which might be left to settle (in the case of medicine, in accordance with the State) the terms on which they grant degrees.

The university or universities should be learning and teaching bodies devoted to art (literary and other), history, philosophy, and science, where any one who wanted to learn all that is known about these matters should find people who could teach him and put him in the way of learning for himself.

That is what the world will want one day or other, as a supplement to all manner of high schools and technical institutions in which young people get decently educated and learn to earn their bread—such as our present universities.

It will be a place for men to get knowledge ; and not for boys and adolescents to get degrees.

I wish I could get the younger men like yourself to see that this is the goal which they may reach, and in the meanwhile to take care that no such Philistine compromise as is possible at present, becomes too strong to survive a sharp shake.—I am, yours very faithfully,

T. H. HUXLEY.

He sketches his ideal of a modern university, and especially of its relation to the Medical Schools, in a letter to Professor Ray Lankester of April 11 :—

HODESLEA, EASTBOURNE,
April 11, 1892.

MY DEAR LANKESTER—We have been having ten days of sunshine, and I have been correspondingly lazy, especially about letter-writing. This, however, is my notion ; that unless people clearly understand that the university of the future is to be a very different thing from the university of the past, they had better put off meddling for another generation.

The mediæval university looked backwards : it professed to be a storehouse of old knowledge, and except in the way of dialectic cobweb-spinning, its professors had nothing to do with novelties. Of the historical and physical (natural) sciences, of criticism and laboratory practice, it knew nothing. Oral teaching was of supreme importance on account of the cost and rarity of manuscripts.

The modern university looks forward, and is a factory of new knowledge : its professors have to be at the top of the wave of progress. Research and criticism must be the breath of their nostrils ; laboratory work the main business of the scientific student ; books his main helpers.

The lecture, however, in the hands of an able man will still have the utmost importance in stimulating and giving facts and principles their proper relative prominence.

I think we should get pretty nearly what is wanted by grafting a Collége de France on to the University of London, subsidising University College and King's College (if it will get rid of its tests, not otherwise), and setting up two or three more such bodies in other parts of London. (Scotland, with a smaller population than London, has four complete universities !)

I should hand over the whole business of medical education and graduation to a medical universitas to be constituted by the royal colleges and medical schools, whose doings, of course, would be checked by the Medical Council.

Our side has been too apt to look upon medical schools as feeders for Science. They have been so, but to their detriment as medical schools. And now that so many opportunities for purely scientific training are afforded, there is no reason they should remain so.

The problem of the Medical University is to make an average man into a good practical doctor before he is twenty-two, and with not more expense than can be afforded by the class from which doctors are recruited, or than will be rewarded by the prospect of an income of £400 to £500 a year.

It is not right to sacrifice such men, and the public on whom they practise, for the prospect of making 1 per cent of medical students into men of science.—Ever yours very faithfully,   T. H. HUXLEY.

An undated draft in his own handwriting (probably the draft of a speech delivered the first time he came to the committee as President, October 26) expands the same idea as to the modern requirements of the University :—

The cardinal fact in the University question appears to me to be this : that the student to whose wants the mediæval University was adjusted, looked to the past and sought book-learning, while the modern looks to the future and seeks the knowledge of things.

The mediæval view was that all knowledge worth having was explicitly or implicitly contained in various ancient writings; in the Scriptures, in the writings of the greater Greeks, and those of the Christian Fathers. Whatever apparent novelty they put forward, was professedly obtained by deduction from ancient data.

The modern knows that the only source of real

knowledge lies in the application of scientific methods of inquiry to the ascertainment of the facts of existence; that the ascertainable is infinitely greater than the ascertained, and that the chief business of the teacher is not so much to make scholars as to train pioneers.

From this point of view, the University occupies a position altogether independent of that of the coping-stone of schools for general education, combined with technical schools of Theology, Law, and Medicine. It is not primarily an institution for testing the work of schoolmasters, or for ascertaining the fitness of young men to be curates, lawyers, or doctors.

It is an institution in which a man who claims to devote himself to Science or Art, should be able to find some one who can teach him what is already known, and train him in the methods of knowing more.

I include under Art,—Literature, the pictorial and plastic art with Architecture, and Music; and under Science,—Logic, Philosophy, Philology, Mathematics, and the Physical Sciences.

The question of the connection of the High Schools for general education, and of the technical schools of Theology, Law, Medicine, Engineering, Art, Music, and so on, with the University is a matter of practical detail. Probably the teaching of the subjects which stand in the relation of preliminaries to technical teaching and final studies in higher general education in the University would be utilised by the colleges and technical schools.

All that I have to say on this subject is, that I see no reason why the existing University of London should not be completed in the sense I have defined by grafting upon it a professoriate with the appropriate means and appliances, which would supply London with the analogue of the Ecole des hautes Études and the Collége de France in Paris, and of the Laboratories with the Professor Extraordinarius and Privat Docenten in the German Universities.

A new Commission was promised to look into the whole question of the London University. This is referred to in a letter to Sir J. Donnelly of March 30, 1892.

Unless you want to kill Foster, don't suggest him for the Commission. He is on one already.

The whole affair is a perfect muddle of competing crude projects and vested interests, and is likely to end in a worse muddle, as anything but a patch up is, I believe, outside practical politics at present.

If I had *carte blanche*, I should cut away the technical "Faculties" of Medicine, Law, and Theology, and set up first-class chairs in Literature, Art, Philosophy, and pure Science—a sort of combination of Sorbonne (without Theology) and Collége de France.

Thank Heaven I have never been asked to say anything, and my chimæras remain *in petto*. They would be scouted.

On the other hand, he was most anxious to keep the School of Science at South Kensington entirely independent. He writes again on May 26 :—

I trust Rücker and Thorpe are convinced by this time that I knew what I was talking about when I told them, months ago, that there would be an effort to hook us into the new University hotch-potch.

I am ready to oppose any such project tooth and nail. I have not been striving these thirty years to get Science clear of their schoolmastering sham-literary peddling to give up the game without a fight. I hope my Lords will be staunch.

I am glad my opinion is already on record.

And similarly to Sir M. Foster on October 30 :—

You will have to come to London and set up
physiology at the Royal College of Science. It is the
only place in Great Britain in which scientific teaching
is trammelled neither by parsons nor by littérateurs. I
have always implored Donnelly to keep us clear of any
connection with a University of any kind, sort, or description, and I tried to instil the same lesson into the doctors
the other day. But the "liberal education" cant is an
obsession of too many of them.

A further step was taken in June, when he was
sent a new draft of proposals, afterwards adopted by
the above-mentioned general meeting of the Association in March 1893, sketching a constitution for a
new university, and asking for the appointment of
a Statutory Commission to carry it out. The University thus constituted was to be governed by a
Court, half of which should consist of university professors [1]; it was to include such faculties as Law,

---

[1] "As for a government by professors only" (he writes in the
Times of Dec. 6, 1892), "the fact of their being specialists is
against them. Most of them are broad-minded, practical men ;
some are good administrators. But, unfortunately, there is among
them, as in other professions, a fair sprinkling of one-idea'd
fanatics, ignorant of the commonest conventions of official relation,
and content with nothing if they cannot get everything their own
way. It is these persons who, with the very highest and purest
intentions, would ruin any administrative body unless they were
counterpoised by non-professional, common-sense members of
recognised weight and authority in the conduct of affairs."
Furthermore, against the adoption of a German university system,
he continues, "In holding up the University of Berlin as our
model, I think you fail to attach sufficient weight to the considerations that there is no Minister of Public Instruction in these
realms ; that a great many of us would rather have no university
at all than one under the control of such a minister, and whose
highest representatives might come to be, not the fittest men, but
those who stood foremost in the good graces of the powers that be,
whether Demos, Ministry, or Sovereign."

Engineering, Medicine, while it was to bring into connection the various teaching bodies scattered over London. The proposers themselves recognised that the scheme was not ideal, but a compromise which at least would not hamper further progress, and would supersede the Gresham scheme, which they regarded as a barrier to all future academic reform.

The Association as thus constituted Huxley now joined, and was immediately asked to accept the Presidency, not that he should do any more militant work than he was disposed to attempt, but simply that he should sit like Moltke in his tent and keep an eye on the campaign.

He felt it almost a point of honour not to refuse his best services to a cause he had always had at heart, though he wrote :—

There are some points in which I go further than your proposals, but they are so much, to my mind, in the right direction that I gladly support them.

And again :—

The Association scheme is undoubtedly a compromise— but it is a compromise which takes us the right way, while the former schemes led nowhere except to chaos.

He writes to Sir W. H. Flower :—

HODESLEA, EASTBOURNE,
June 27, 1892.

MY DEAR FLOWER—I had quite given up the hope that anything but some wretched compromise would come of the University Commission, when I found, to my surprise, no less than gratification, that a strong party among the younger men were vigorously taking the matter up in the right (that is, *my*) sense.

In spite of all my good resolves to be a " hermit old in mossy cell," I have enlisted—for ambulance service if nothing better.

The move is too important to spare oneself if one can be of any good.—Ever yours very faithfully,

T. H. HUXLEY.

Of his work in this position Professor Karl Pearson says, in a letter to me :—

Professor Huxley gallantly came to lead a somewhat forlorn hope,—that of establishing a really great university in London. He worked, as may naturally be supposed, with energy and persistence, and one, who like myself was not in full sympathy with the lines he took, can but admire the vigour he threw into the movement. Nothing came of it practically ;... but Professor Huxley's leadership did, at any rate, a great deal to unite the London teachers, and raise their ideal of a true university, while at the same time helping to repress the self-interests of many persons and institutions which had been before very much to the front.

Clearly this is the sort of thing referred to in a letter of December 20 :—

Got through the Association business very well, but had to show that I am the kind of head that does not lend itself to wagging by the tail.

The Senate of the University of London showed practical unanimity in accepting the idea of taking on teaching functions if the Commission should think it desirable, though the Medical Schools were still desirous of getting their degree granted on the mere license examination of the Royal Colleges, without any evidence of general culture or academical training, and on July 28 Huxley writes :—

The decision of the representatives of the Medical Schools is just such as I should have expected.   I always told my colleagues in the Senate of the University of London that such was their view, and that, in the words of Pears' advertisement, they "would not be happy till they got it."

And they won't get it unless the medical examining bodies are connected into a distinct degree-giving body.

In the course of the autumn matters seemed to be progressing.   He writes to Sir M. Foster, November 9 :—

I am delighted to say that Paget [1] has taken up the game, and I am going to a committee of the University this day week to try my powers of persuasion.   If the Senate can only be got to see where salvation lies and strike hard without any fooling over details, we shall do a great stroke of business for the future generations of Londoners.

And by the end of the year he writes :—

I think we are going to get something done, as the Senate of the U.L. has come into line with us, and I hope University College will do the same.

Meanwhile he was asked if he would appear before the Commission and give evidence—to " talk without interrogation " so as to convince the Commission of the inadequacy of the teaching of science in general

Sir James Paget, Vice-Chancellor of the University.

and of the absence of means and appliances for the
higher teaching. This he did early in January 1893,
representing partly his own views, partly those of
the Association, to whom he read what he proposed
to say, before being authorised to speak on their
behalf.

His position is finally defined by the following
letter :—

Feb. 9, 1893.

DEAR PROFESSOR WELDON—I wish anything I have
said or shall say about the organisation of the New
University to be taken in connection with the following
postulates which I conceive to be of primary importance :

1. The New University is not to be a separate body
from the present University of London.

2. All persons giving academic instruction of a certain
rank are to be " University Professors."

3. The Senate is to contain a large proportion of
representatives of the " University Professors " with a
limited term of office (say five years).

4. The University chest is to receive all fees and other
funds for University purposes ; and the Professors are to
be paid out of it, according to work done for the
University — thus putting an end to the present commercial competition of teaching institutions.

5. In all questions of Teaching, Examination, and
Discipline the authority of the Senate is to be supreme—
(saving appeal to the Privy Council).

Your questions will be readily answered if these
postulates are kept in view.

In the case you put, the temptation to rivalry would
not exist ; and I should imagine that the Senate would
refuse funds for the purpose of duplicating an existing
Institution, unless very strong grounds for so doing could

be shown. In short, they would adopt the plan which commends itself to you.

That to which I am utterly opposed is the creation of an Established Church Scientific, with a hierarchical organisation and a professorial Episcopate. I am fully agreed with you that all trading competition between different teaching institutions is a thing to be abolished (see No. 4 above).

On the other hand, intellectual competition is a very good thing, and perfect freedom of learning and teaching the best of all things.

If you put a physical, chemical, or biological bishop at the head of the teachers of those sciences in London, you will do your best to destroy that freedom. My bar to any catastrophe of that sort lies in No. 3. Let us take the case of Biology. I suppose there will be, at least, half a dozen Professoriates in different branches of this subject; each Professor will be giving the same amount of time and energy to University work, and will deserve the same pay. Each, if he is worth his salt, will be a man holding his own views on general questions, and having as good a right as any other to be heard. Why is one to be given a higher rank and vastly greater practical influence than all the rest? Why should not each be a "University Professor" and have his turn on the Senate in influencing the general policy of the University? The nature of things drives men more and more into the position of specialists. Why should one specialist represent a whole branch of science better than another, in Council or in Administration?

I am afraid we cannot build upon the analogy of Cambridge. In the first place, London is not Cambridge; and, in the second, Michael Fosters do not grow on every bush.

The besetting sin of able men is impatience of contradiction and of criticism. Even those who do their best to resist the temptation, yield to it almost un-

consciously and become the tools of toadies and flatterers.
" Authorities," " disciples," and " schools " are the curse of
science ; and do more to interfere with the work of the
scientific spirit than all its enemies.

Thus you will understand why I have so strongly
opposed " absorption."   No one can feel more strongly
than I the need of getting the present chaos into order
and putting an end to the absurd waste of money and
energy.   But I believe that end may be attained by the
method of unification which I have suggested ; without
bringing in its train the evils which will inevitably flow
from " absorptive " regimentation.

What I want to see is such an organisation of the
means and appliances of University instruction in all its
branches, as will conduce to the largest possible freedom
of research, learning, and teaching.   And if anybody will
show me a better way to that end than through the
measures I have suggested, I will gladly leave all and
follow him.—I am yours very faithfully,

T. H. Huxley.

P.S.—Will you be so kind as to let Professor Lankester
see this letter, as I am writing to him and shirk the
labour of going over the whole ground again.

His last public activity, indeed, was on behalf of
University reform, when in January 1895 he
represented not only the Association, but, in the
enforced absence of Sir James Paget, the Senate of
the University also, on a deputation to Lord Rosebery,
then Prime Minister, to whom he wrote asking if he
were willing to receive such a deputation.

Hodeslea, Eastbourne,
Dec. 4, 1894.

Dear Lord Rosebery—A number of scientific
people, in fact I think I may say all the leading men of

science, and especially teachers in the country, are very anxious to see the University of London reorganised upon the general principles set forth in the Report of the last Royal Commission.

To this end nothing is wanted but the institution of a strong Statutory Commission ; and we have all been hoping that a Bill would be introduced for that purpose.

It is rumoured that there are lions in the path. But even lions are occasionally induced to retreat by the sight of a large body of beaters. And some of us think that such a deputation as would willingly wait on you, might hasten the desired movement.

We proposed something of the kind to Mr. Acland months ago, but nothing has come of the suggestion—not, I am sure, from any want of good will to our cause on his part.

Within the last few days I have been so strongly urged to bring the matter before you, that in spite of some doubts as to the propriety of going beyond my immediate chief the V.P.[1] even in my private capacity I venture to make this appeal.—I am, dear Lord Rosebery, faithfully yours, THOS. H. HUXLEY.

---

[1] The Vice-President of the Committee of Council, Mr. Acland.

CHAPTER X

1892

SEVERAL letters of this year touch on educational subjects. The following advice as to the best training for a boy in science, was addressed to Mr. Briton Riviere, R.A. :—

HODESLEA, *June* 19, 1892.

MY DEAR RIVIERE—Touching the training of your boy who wants to go in for science, I expect you will have to make a compromise between that which is theoretically desirable and that which is practically most advantageous, things being as they are.

Though I say it that shouldn't, I don't believe there is so good a training in physical science to be got anywhere as in our College at South Kensington. But Bernard could hardly with advantage take this up until he is seventeen at least. What he would profit by most as a preliminary, is training in the habit of expressing himself well and clearly in English ; training in mathematics and the elements of physical science ; in French and German, so as to read those languages easily—especially German ; in drawing—not for hifalutin art, of which he will probably have enough in the blood—but accurate dry reproduction of form—one of the best disciplines of the powers of observation extant.

On the other hand, in the way of practical advantage

in any career, there is a great deal to be said for sending a clever boy to Oxford or Cambridge. There are not only the exhibitions and scholarships, but there is the rubbing shoulders with the coming generation which puts a man in touch with his contemporaries as hardly anything else can do. A very good scientific education is to be had at both Cambridge and Oxford, especially Cambridge now.

In the case of sending to the university, putting through the Latin and Greek mill will be indispensable. And if he is not going to make the classics a serious study, there will be a serious waste of time and energy.

So much in all these matters depends on the x contained in the boy himself. If he has the physical and mental energy to make a mark in science, I should drive him straight at science, taking care that he got a literary training through English, French, and German. An average capacity, on the other hand, may be immensely helped by university means of flotation.

But who in the world is to say how the x will turn out, before the real strain begins? One might as well prophesy the effect of a glass of "hot-with" when the relative quantities of brandy, water, and sugar are unknown. I am sure the large quantity of brandy and the very small quantity of sugar in my composition were suspected neither by myself, nor any one else, until the rows into which wicked men persisted in involving me began!

And that reminds me that I forgot to tell the publishers to send you a copy of my last peace-offering,[1] and that one will be sent you by to-morrow's post. There is nothing new except the prologue, the sweet reasonableness of which will, I hope, meet your approbation.

It is not my fault if you have had to toil through this frightfully long screed; Mrs. Riviere, to whom our love, said you wanted it. "Tu l'as voulu, Georges Dandin."— Ever yours very faithfully,   T. H. HUXLEY.

---

[1] The *Essays on Controverted Questions.*

The following deals with State intervention in intermediate education :—

(For Sunday morning's leisure, or take it to church and read it in your hat.)

HODESLEA, EASTBOURNE,
Oct. 1, 1892.

MY DEAR DONNELLY—Best thanks for sending on my letter. I do not suppose it will do much good, but, at any rate, I thought I ought to try to prevent their making a mess of medical education.

I like what I have seen of Acland. He seemed to have both intelligence and volition.

As to intermediate education I have never favoured the notion of State intervention in this direction.

I think there are only two valid grounds for State meddling with education : the one the danger to the community which arises from dense ignorance ; the other, the advantage to the community of giving capable men the chance of utilising their capacity.

The first furnishes the justification for compulsory elementary education. If a child is taught reading, writing, drawing, and handiwork of some kind ; the elements of mathematics, physics, and history, and I should add of political economy and geography ; books will furnish him with everything he can possibly need to make him a competent citizen in any rank of life.

If with such a start, he has not the capacity to get all he needs out of books, let him stop where he is. Blow him up with intermediate education as much as you like, you will only do the fellow a mischief and lift him into a place for which he has no real qualification. People never will recollect, that mere learning and mere cleverness are of next to no value in life, while energy and intellectual grip, the things that are inborn and cannot be taught, are everything.

The technical education act goes a long way to meet the second claim of the State; so far as scientific and industrial capacities are concerned. In a few years there will be no reason why any potential Whitworth or Faraday, in the three kingdoms, should not readily obtain the best education that is to be had, scientific or technical. The same will hold good for Art. So the question that arises seems to me to be whether the State ought or ought not to do something of the same kind for Literature, Philosophy, History, and Philology.

I am inclined to think not, on the ground that the universities and public schools ought to do this very work, and that as soon as they cease to be clericalised seminaries they probably will do it.

If the present government would only give up their Irish fad—and bring in a bill to make it penal for any parson to hold any office in a public school or university or to presume to teach outside the pulpit—they should have my valuable support!

I should not wonder if Gladstone's mind is open on the subject. Pity I am not sufficiently a *persona grata* with him to offer to go to Hawarden and discuss it.

I quite agree with you, therefore, that it will play the deuce if intermediate education is fossilised as it would be by any Act prepared under present influences. The most I should like to see done, would be to help the youth of special literary, linguistic and so forth, capacity, to get the best training in their special line.

It was lucky we did not go to you. My wife got an awful dose of neuralgia and general upset, and was laid up at the Hotel. The house was not quite finished inside, but we came in on Tuesday, and she has been getting better ever since in spite of the gale.

I am sorry to hear of the recurrence of influenza. It is a beastly thing. Lord Justice Bowen told me he has had it every time it has been in the country. You must

come and try Eastbourne air as soon as we are settled.
With our love to you and Mrs. Donnelly—Ever yours,
T. H. HUXLEY.

Better be careful, I return all letters on which R.H.[1]
is not in full.

The next is to a young man with aspirations after
an intellectual career, who asked his advice as to the
propriety of throwing up his business, and plunging
into literature or science :—

HODESLEA, EASTBOURNE,
Nov. 5, 1892.

DEAR SIR—I am very sorry that the pressure of other
occupations has prevented me from sending an earlier
reply to your letter.

In my opinion a man's first duty is to find a way of
supporting himself, thereby relieving other people of the
necessity of supporting him.   Moreover, the learning to
do work of practical value in the world, in an exact and
careful manner, is of itself a very important education,
the effects of which make themselves felt in all other
pursuits.   The habit of doing that which you do not care
about when you would much rather be doing something
else, is invaluable.   It would have saved me a frightful
waste of time if I had ever had it drilled into me in
youth.

Success in any scientific career requires an unusual
equipment of capacity, industry, and energy.   If you
possess that equipment you will find leisure enough after
your daily commercial work is over, to make an opening
in the scientific ranks for yourself.   If you do not, you
had better stick to commerce.   Nothing is less to be
desired than the fate of a young man, who, as the Scotch
proverb says, in " trying to make a spoon spoils a horn,"

---

[1] An allusion to his recent Privy Councillorship   See p. 247.

and becomes a mere hanger-on in literature or in science,
when he might have been a useful and a valuable member
of Society in. other occupations.

I think that your father ought to see this letter.—
Yours faithfully,   T. H. HUXLEY.

The last of the series, addressed to the secretary
of a free-thought association, expresses his firmly
rooted disgust at the use of mere ribaldry in attacking the theological husks which enclose a religious
ideal.

May 22, 1892.

DEAR SIR—I regret that I am unable to comply with
the wish of your committee. For one thing, I am
engaged in work which I do not care to interrupt, and
for another, I always make it a rule in these matters to
" fight for my own hand." I do not desire that any one
should share my responsibility for what I think fit to
say, and I do not wish to be responsible for the opinions
and modes of expression of other persons.

I do not say this with any reference to Mr. ——
who is a sober and careful writer. But both as a matter
of principle and one of policy, I strongly demur to a
great deal of what appears as " free thought " literature,
and I object to be in any way connected with it.
Heterodox ribaldry disgusts me, I confess, rather more
than orthodox fanaticism. It is at once so easy; so
stupid; such a complete anachronism in England, and
so thoroughly calculated to disgust and repel the very
thoughtful and serious people whom it ought to be the
great aim to attract. Old Noll knew what he was about
when he said that it was of no use to try to fight the
gentlemen of England with tapsters and serving-men.
It is quite as hopeless to fight Christianity with scurrility.
We want a regiment of Ironsides.

This summer brought Huxley a most unexpected distinction in the shape of admission to the Privy Council.   Mention has already been made (vol. ii. p. 56) of his reasons for refusing to accept a title for distinction in science, apart from departmental administration.   The proper recognition of science, he maintained, lay in the professional recognition of a man's work by his peers in science, the members of the learned societies of his own and other countries.

But, as has been said, the Privy Councillorship was an office, not a title, although with a title attaching to the office; and in theory, at least, a scientific Privy Councillor might some day play an important part as an accredited representative of science, to be consulted officially by the Government, should occasion arise.

Of a selection of letters on the subject, mostly answers to congratulations, I place first the one to Sir M. Foster, which gives the fullest account of the affair.

CORS-Y-GEDOL HOTEL, BARMOUTH,
Aug. 23, 1892.

MY DEAR FOSTER—I am very glad you think I have done rightly about the P.C.; but in fact I could hardly help myself.

Years and years ago I was talking to Donnelly about these things, and told him that so far as myself was concerned, I would have nothing to do with official decorations—didn't object to other people having them, especially heads of offices, like Hooker and Flower—but preferred to keep clear myself.   But I added that there was one thing I did not mind telling him, because no

English Government would ever act upon my opinion—
and that was that the P.C. was a fit and proper recognition for science and letters.   I have no doubt that he
has kept this in mind ever since—in fact Lord Salisbury's
letter (which was very handsome) showed he had been
told of my *obiter dictum*.   Donnelly was the first channel
of inquiry whether I would accept, and was very strong
that I should.

So you see if I had wished to refuse it, it would have
been difficult and ungracious.   But, on the whole, I thought
the precedent good.   Playfair tells me he tried to get it
done in the case of Faraday and Babbage thirty years ago,
and the thing broke down.   Moreover a wicked sense of
the comedy of advancing such a pernicious heretic, helped
a good deal.

The worst of it is, I have just had a summons to go to
Osborne on Thursday and it is as much as I shall be able
to do.

We have been in South Wales, in the neighbourhood
of the Colliers, and are on our way to the Wallers for the
Festival week at Gloucester.   We hope to get back to
Eastbourne in the latter half of September and find the
house clean swept and garnished.   After that, by the way,
it is *not* nice to say that we shall hope to have a visit
from Mrs. Foster and you.

With our love to you both—Ever yours,

T. H. Huxley.

I am glad you are resting, but oh, why another
Congress !

Hodeslea, Eastbourne,
June 21, 1892.

My dear Donnelly—You have been and done me at
last, you betrayer of confidence.   This is what comes of
confiding one's pet weakness to a bosom-friend !

But I can't deny my own words, or the accuracy of
your devil of a memory—and, moreover, I think the
precedent of great importance.

I have always been dead against orders of merit and the like, but I think that men of letters and science who have been of use to the nation (Lord knows if I have) may fairly be ranked among its nominal or actual councillors.

As for yourself, it is only one more kindness on the top of a heap so big I shall say nothing about it.

Mrs. Right Honourable sends her love to you both, and promises not to be proud.—Ever yours very faithfully,

T. H. HUXLEY.

CORS-Y-GEDOL HOTEL, BARMOUTH,
Aug. 20, 1892.

MY DEAR DONNELLY—I began to think that Lord Salisbury had thought better of it—(I should not have been surprised at all if he had) and was going to leave me a P.P.C. instead of a P.C. when the announcement appeared yesterday.

This morning, however, I received his own letter (dated the 16th), which had been following me about. A very nice letter it is too—he does the thing handsomely while he is about it.

Well, I think the thing is good for science; I am not such a self humbug as to pretend that my vanity is not pleasantly tickled; but I do not think there is any aspect of the affair more pleasant to me, than the evidence it affords of the strength of our old friendship. Because with all respect for my noble friends, deuce a one would ever have thought of it, unless you had not only put it—but rubbed it—into their heads.

I have not forgotten that private and confidential document that you were so disgusted to find had been delivered to me ! You have tried it on before—so don't deny it.

But bless my soul, how profound is old Cole's remark about the humour of public affairs. To think of a Conservative Government—pride of the Church—going out of its way to honour one not only of the wicked, but of the notoriousest and plain-spoken wickedness. My wife and I drove over to Dolgelly yesterday—do you know it? one of the loveliest things in the three kingdoms—and every now and then had a laugh over this very quaint aspect of the affair.

Can you tell me what I shall have to do in the dim and distant future? I suppose I shall have to go and swear somewhere (I am always ready to do that on occasion). Is admission to the awful presence of H.M. involved? Shall I have to rig up again in that Court suit, which I hoped was permanently laid up in lavender? Resolve me these things.

We shall be here I expect at least another week; and bring up at Gloucester about the 3rd September. Hope to get back to Hodeslea latter part of September.—Ever yours faithfully.   T. H. HUXLEY.

TO SIR J. D. HOOKER

Aug. 20.

You will have seen that I have been made a P.C. If I had been offered to be made a police constable I could not have been more flabbergasted than I was when the proposition came to me a few weeks ago. I will tell you the story of how it all came about when we meet. The Archbishopric of Canterbury is the only object of ambition that remains to me. Come and be Suffragan; there is plenty of room at Lambeth and a capital garden!

TO HIS YOUNGEST DAUGHTER

CORS-Y-GEDOL HOTEL, BARMOUTH,
Aug. 22, 1892.

DEAREST BABS—If Lord Salisbury had known my address, M——-and I should have had our little joke out before leaving Saundersfoot,[1] as the letter was dated 16th. It must be a month since Lord Cranbrook desired Donnelly to find out if I would accept the P.C., and as I heard no more about it up to the time of dissolution, I imagined there was a hitch somewhere. And really, the more I think of it the queerer does it seem, that a Tory and Church Government should have delighted to honour the worst-famed heretic in the three kingdoms.

I am sure Donnelly has been at the bottom of it, as he is the only person to whom I ever spoke of the fitness of the P.C. for men of science and letters.

The queer thing is that his chief and Lord Salisbury listened to the suggestion.

Tell Jack he is simply snuffed out—younger sons of peers go with the herd of Barts. and knights, I believe. But a table of precedence is not to be had for love or money—and my anxiety is wearing.

This place is as perfectly delightful as Aberystwith was t'other. . . . With best love to you all—Ever your

PATER.

TO MRS. W. K. CLIFFORD

CORS-Y-GEDOL HOTEL, BARMOUTH,
Aug. 22, 1892.

MY DEAR LUCY—I am glad to think that it is the honours that blush and not the recipient, for I am past that form of vascular congestion.

It was known that the only peerage I would accept was a spiritual one ; and as H.M. shares the not unnatural prejudice which led her illustrious predecessor (now some time dead) to object to give a bishopric to Dean Swift, it was thought she could not stand the promotion of Dean Huxley ; would see * him in fact . . .   * This is a pun.

---

[1] Where he had been staying with his daughter.

Lord S. apologised for not pressing the matter, but pointed out that, as Evolutionism is rapidly gaining ground among the people who have votes, it was probable, if not certain, that his eminent successor (whose mind is always open) would become a hot evolutionist before the expiration of the eight months' office which Lord S. (who needs rest) means to allow him. And when eminent successor goes out, my bishopric will be among the Dissolution Honours. If H.M. objects she will be threatened with the immediate abolition of the H. of Lords, and the institution of a social democratic federation of counties, each with an army, navy, and diplomatic service of its own.

I know you like to have the latest accurate intelligence, but this really must be considered confidential. As a P.C. I might lose my head for letting out State secrets.— Ever your affectionate   PATER.

TO SIR JOSEPH FAYRER

CORS-Y-GEDOL HOTEL, BARMOUTH, WALES,
Aug. 28, 1892.

It is very pleasant to get the congratulations of an old friend like yourself. As we went to Osborne the other day I looked at the old *Victory* and remembered that six and forty years ago I went up her side to report myself on appointment, as a poor devil of an assistant surgeon. And I should not have got that far if you had not put it into my head to apply to Burnett.

TO SIR JOSEPH PRESTWICH

CORS-Y-GEDOL HOTEL, BARMOUTH,
Aug. 31, 1892.

MY DEAR PRESTWICH—Best thanks for your congratulations. As I have certainly got more than my temporal

deserts, the other " half " you speak of can be nothing less
than a bishopric ! May you live to see that dignity conferred ; and go on writing such capital papers as the last
you sent me, until I write myself your Right Revd. as
well as Right Honble. old friend,   T. H. HUXLEY.

To Sir W. H. Flower

CORS-Y-GEDOL HOTEL, BARMOUTH,
Aug. 31, 1892.

MY DEAR FLOWER—Many thanks for your congratulations, with Lady Flower's postscript not forgotten.   I
should have answered your letter sooner, but I had to go
to Osborne last week in a hurry, kiss hands and do my
swearing.   It was very funny that the Gladstone P.C.'s
had the pleasure of welcoming the Salisbury P.C.'s among
their first official acts !

I will gladly come to as many meetings of the Trustees
as I can.   Only you must not expect me in very severe
weather like that so common last year.   My first attack
of pleurisy was dangerous and not painful ; the second
was painful and not dangerous ; the third will probably
be both painful and dangerous, and my commander-in-
chief (who has a right to be heard in such matters) will
not let me run the risk of it.

But I have marked down Oct. 22 and Nov. 24, and
nothing short of snow shall stop me.

As to what you want to do, getting butter out of a
dog's mouth is an easier job than getting patronage out
of that of a lawyer or an ecclesiastic.   But I am always
good for a forlorn hope, and we will have a try.

We shall not be back at Eastbourne till the latter
half of September, and I doubt if we shall get into our
house even then.   We leave this for Gloucester, where we
are going to spend the festival week with my daughter
to-morrow.—With our love to you both, ever yours very
faithfully,   T. H. HUXLEY.

I see a report that Owen is sinking. Poor old man; it seems queer that just as I am hoist to the top of my tree he should be going underground. But at 88 life cannot be worth much.

To Mr. W. F. Collier

Cors-y-Gedol Hotel, Barmouth Water,
Aug. 31, 1892.

Accept my wife's and my hearty thanks for your kind congratulations. When I was a mere boy I took for motto of an essay, "What is honour? Who hath it? He that died o' Wednesday," and although I have my full share of ambition and vanity, I doubt not, yet Falstaff's philosophical observation has dominated my mind and acted as a sort of perpetual refrigerator to these passions. So I have gone my own way, sought for none of these things and expected none—and it would seem that the deepest schemer's policy could not have answered better. We must have a new Beatitude, "Blessed is the man who expecteth nothing," without its ordinary appendix.

I tell Jack [1] I have worked hard for a dignity which will enable me to put down his aristocratic swaggering.

It took some time, however, to get used to the title, and it was October before he wrote :—

The feeling that "The Right Honble." on my letters is a piece of chaff is wearing off, and I hope to get used to my appendix in time.

The "very quaint" ceremony of kissing hands is described at some length in a letter to Mrs. Huxley from London on his way back from Osborne :—

[1] His son-in-law, Hon. John Collier.

GREAT WESTERN HOTEL,
Aug. 25, 1892, 6.40 P.M.

I have just got back from Osborne, and I find there are a few minutes to send you a letter—by the help of the extra halfpenny. First-rate weather there and back, a special train, carriage with postillions at the Osborne landing-place, and a grand procession of officers of the new household and P.C.'s therein. Then waiting about while the various "sticks" were delivered.

Then we were shown into the presence chamber where the Queen sat at a table. We knelt as if we were going to say our prayers, holding a testament between two, while the Clerk of the Council read an oath of which I heard not a word. We each advanced to the Queen, knelt and kissed her hand, retired backwards, and got sworn over again (Lord knows what I promised and vowed this time also). Then we shook hands with all the P.C.'s present, including Lord Lorne, and so exit backwards. It was all very curious. . . .

After that a capital ·lunch and back we came. Ribblesdale and several other people I knew were of the party, and I found it very pleasant talking with him and Jesse Collings, who is a very interesting man.

"Oh," he said, "how I wish my poor mother, who was a labouring woman—a great noble woman—and brought us nine all up in right ways, could have been alive." Very human and good and dignified too, I thought.

He also used to tell how he was caught out when he thought to make use of the opportunity to secure a close view of the Queen. Looking up, he found her eyes fixed upon him ; Her Majesty had clearly taken the opportunity to do the same by him.

Regarding the Privy Councillorship as an exceptional honour for science, over and above any recognition of his personal services, which he thought amply met by the Civil List pension specially con- ferred upon him as an honour at his retirement from the public service, Huxley was no little vexed at an article in *Nature* for August 25 (vol. xlvi. p. 397), reproaching the Government for allowing him to leave the public service six years before, without recognition. Accordingly he wrote to Sir J. Donnelly on August 27 :—

It is very unfair to both Liberal and Conservative Governments, who did much more for me than I expected, and I feel that I ought to contradict the statement without loss of time.

So I have written the inclosed letter for publication in *Nature.* But as it is always a delicate business to meddle with official matters, I wish you would see if I have said anything more than I ought to say in the latter half of the letter. If so, please strike it out, and let the first half go.

I had a narrow shave to get down to Osborne and kiss hands on Thursday. What a quaint ceremony it is!

The humour of the situation was that we three hot Unionists, White Ridley, Jesse Collings, and I, were escorted by the whole Gladstonian household.

And again on August 30 :—

In the interview I had with Lord Salisbury on the subject of an order of merit—ages ago [1]—I expressly gave him to understand that I considered myself out of the running—having already received more than I had any right to expect. And when he has gone out of his

---

[1] See p. 23.

way to do honour to science, it is stupid of *Nature* to strike the discordant note.

His letter appeared in *Nature* of September 1 (vol. xlvi. p. 416). In it he declared that both Lord Salisbury's and Mr. Gladstone's Governments had given him substantial recognition; that Lord Iddesleigh had put the Civil List pension expressly as an honour; and finally, that he himself placed this last honour in the category of "unearned increments."

CHAPTER XI

1892

THE following letters are mainly of personal interest;
some merely illustrate the humorous turn he would
give to his more intimate correspondence; others
strike a more serious note, especially those to friends
whose powers were threatened by overwork or ill-
health.

With these may fitly come two other letters; one
to a friend on his re-marriage, the other to his
daughter, in reply to a birthday letter.

My wife and I send our warmest good wishes to your
future wife and yourself. I cannot but think that those
who are parted from us, if they have cognisance of what
goes on in this world, must rejoice over everything that
renders life better and brighter for the sojourners in it—
especially of those who are dear to them. At least, that
would be my feeling.

Please commend us to Miss ——, and beg her not to
put us on the " Index," because we count ourselves among
your oldest and warmest friends.

To his Daughter, Mrs. Roller

HODESLEA, EASTBOURNE,
May 5, 1892.

It was very pleasant to get your birthday letter and the photograph, which is charming.

The love you children show us, warms our old age better than the sun.

For myself, the sting of remembering troops of follies and errors, is best alleviated by the thought that they may make me better able to help those who have to go through like experiences, and who are so dear to me that I would willingly pay an even heavier price, to be of use. Depend upon it, that confounded "just man who needed no repentance" was a very poor sort of a father. But perhaps his daughters were "just women" of the same type; and the family circle as warm as the interior of an ice-pail.

A certain artist, who wanted to have Huxley sit to him, tried to manage the matter through his son-in-law, Hon. J. Collier, to whom the following is addressed :—

HODESLEA, EASTBOURNE,
Jan. 27, 1892.

MY DEAR JACK—Inclosed is a letter for you. Will you commit the indiscretion of sending it on to Mr. A. B. if you see no reason to the contrary ?

I hope the subsequent proceedings will interest you no more.

I am sorry you have been so bothered by the critter— but in point of pertinacity he has met his match. (I have no objection to your saying that your father-in-law is a brute, if you think that will soften his disappointment.)

Here the weather has been tropical.   The bananas in the new garden are nearly ripe, and the cocoanuts are coming on.   But of course you expect this, for if it is unbearably sunny in London what must it be here?

All our loves to all of you.—Ever yours affectionately,

PATER.

HODESLEA, EASTBOURNE,
Feb. 1, 1892.

MY DEAR HOOKER—I hear you have influenza rampaging about the Camp ;[1] and I want to point out to you that if you want a regular bad bout of it, the best thing you can do is to go home next Thursday evening, at ten o'clock at night, and plunge into the thick of the microbes, tired and chilled.

If you don't get it then, you will, at any rate, have the satisfaction of feeling that you have done your best!

I am going to the x, but then you see I fly straight after dinner to Collier's per cab, and there is no particular microbe army in Eton Avenue lying in wait for me.

Either let me see after the dinner, or sleep in town, and don't worry.—Yours affectionately,

T. H. HUXLEY.

HODESLEA, EASTBOURNE,
Feb. 19, 1892.

MY DEAR HOOKER—I have just received a notice that Hirst's funeral is to-morrow.   But we are in the midst of the bitterest easterly gale and snowfall we have had all the winter, and there is no sign of the weather mending.

Neither you nor I have any business to commit suicide for that which after all is a mere sign of the affection we have no need to prove for our dear old friend, and the chances are that half an hour cold chapel and grave-side on a day like this would finish us.

---

[1] The name of Sir J. Hooker's house at Sunningdale.

I write this not that I imagine you would think of going, but because my last note spoke so decidedly of my own intention.

But who could have anticipated this sudden reversion to Arctic conditions?—Ever yours affectionately,

T. H. HUXLEY.

HODESLEA, EASTBOURNE,
March 18, 1892.

MY DEAR DONNELLY—My wife got better and was out for a while yesterday, but she is knocked up again to-day.

It would have been very pleasant to see you both, but you must not come down till we get fixed with a new cook and maid, as I believe we are to be in a week or so. None of your hotel-going!

I mourn over the departure of the present cookie—I believe she is going for no other reason than that she is afraid the house will fall on such ungodly people as we are, and involve her in the ruins. That is the modern martyrdom—you don't roast infidels, but people who can roast go to the pious.

Lovely day to-day, nothing but east wind to remind one it is not summer.—Crocuses coming out at last.— Ever yours very faithfully, T. H. HUXLEY.

HODESLEA, EASTBOURNE,
March 27, 1892.

MY DEAR HOOKER—I had to run up to town on Friday and forgot your letter. The x is a puzzle—I will stick by the ship as long as you do, depend upon that. I fear we can hardly expect to see dear old Tyndall there again. As for myself, I dare not venture when snow is on the ground, as on the last two occasions. And now, I am sorry to say, there is another possible impediment in my wife's state of health.

I have had a very anxious time of it altogether lately. But sich is life!

My sagacious grand-daughter Joyce (gone home now) observed to her grandmother some time ago—" I don't want to grow up." " Why don't you want to grow up ? " " Because I notice that grown-up people have a great deal of trouble." Sagacious philosopheress of 7 !—Ever yours affectionately,　　　　　　　　　　　　　T. H. HUXLEY.

HODESLEA, EASTBOURNE,
April 3, 1892.

MY DEAR HOOKER—As I so often tell my wife, " your confounded sense of duty will be the ruin of you." You really, club or no club, had no business to be travelling in such a bitter east wind. However, I hope the recent sunshine has set you up again.

Barring snow or any other catastrophe, I will be at " the Club " dinner on the 26th and help elect the P.R.S. I don't think I go more than once a year, and like you I find the smaller the pleasanter meetings.

I was very sorry to see Bowman's death. What a first-rate man of science he would have been if the Professorship at King's College had been £1000 a year. But it was mere starvation when he held it.

I am glad to say that my wife is much better—thank yours for her very kind sympathy. I was very down the last time I wrote to you.—Ever yours affectionately,

T. H. HUXLEY.

HODESLEA, *June* 27, 1892.

MY DEAR FOSTER—My wife has been writing to Mrs. Foster to arrange for your visit, which will be heartily welcome.

Now I don't want to croak. No one knows better than I, the fatal necessity for any one in your position : more than that, the duty in many cases of plunging into public functions, and all the guttle, guzzle, and gammon therewith connected.

But do let me hold myself up as the horrid example of what comes of that sort of thing for men who have to work as you are doing and I have done. To be sure you are a "lungy" man and I am a "livery" man, so that your chances of escaping candle-snuff accumulations with melancholic prostration are much better. Nevertheless take care. The pitcher is a very valuable piece of crockery, and I don't want to live to see it cracked by going to the well once too often.

I am in great spirits about the new University movement, and have told the rising generation that this old hulk is ready to be towed out into line of battle, if they think fit, which is more commendable to my public spirit than my prudence.—Ever yours,   T. H. HUXLEY.

HODESLEA, *June* 20, 1892.

MY DEAR ROMANES—My wife and I, no less than the Hookers who have been paying us a short visit, were very much grieved to hear that such a serious trouble has befallen you.

In such cases as yours (as I am sure your doctors have told you) hygienic conditions are everything—good air and idleness, *construed strictly,* among the chief. You should do as I have done—set up a garden and water it yourself for two hours every day, besides pottering about to see how things grow (or don't grow this weather) for a couple more.

Sundry box-trees, the majority of which have been getting browner every day since I planted them three months ago, have interested me almost as much as the general election. They typify the Empire with the G.O.M. at work at the root of it!—Ever yours very faithfully,   T. H. HUXLEY.

HODESLEA, *Oct.* 18, 1892.

MY DEAR ROMANES—I throw dust and ashes on my head for having left your letter almost a week unanswered.

But I went to Tennyson's funeral ; and since then my whole mind has been given to finishing the reply forced upon me by Harrison's article in the *Fortnightly*, and I have let correspondence slide.   I think it will entertain you when it appears in November—and perhaps interest— by the adumbration of the line I mean to take if ever that " Romanes " Lecture at Oxford comes off.

As to Madeira—I do not think you could do better. You can have as much quiet there as in Venice, for there are next to no carts or carriages.   I was at an excellent hotel, the " Bona Vista," kept by an Englishman in excellent order, and delightfully situated on the heights outside Funchal.   When once acclimatised and able to bear moderate fatigue, I should say nothing would be more delightful and invigorating than to take tents and make the round of the island.   There is nothing I have seen anywhere which surpasses the cliff scenery of the north side, or on the way thither, the forest of heaths as big as sycamores.

There is a matter of natural history which might occupy without fatiguing you, and especially without calling for any great use of the eyes.   That is the effect of Madeiran climate on English plants transported there —and the way in which the latter are beating the natives.   There is a Doctor who has lots of information on the topic.   You may trust anything but his physic.

[The rest of the letter gives details about scientific literature touching Madeira.]

A piece of advice to his son anent building a house :—

Sept. 22, 1892.

Lastly and biggestly, don't promise anything, agree to anything, nor sign anything (swear you are an " illiterate voter " rather than this last) without advice—or you may find yourself in a legal quagmire. Builders, as a rule, are on a level with horse-dealers in point of honesty—I could tell you some pretty stories from my small experience of them.

The next, to Lord Farrer, is *apropos* of quite an extensive correspondence in the *Times* as to the correct reading of the well-known lines about the missionary and the cassowary, to which both Huxley and Lord Farrer had contributed their own reminiscences.

HODESLEA, *Oct.* 15, 1892.

MY DEAR FARRER—

If *you* were a missionary
In the heat of Timbuctoo
You 'd wear nought but a nice and airy
'Pair of bands—p'raps cassock too.

Don't you see the fine touch of local colour in my version ! Is it not obvious to everybody who understands the methods of high *a priori* criticism that this consideration entirely outweighs the merely empirical fact that your version dates back to 1837—which I must admit is before my adolescence ? It is obvious to the meanest capacity that mine must be the original text in " Idee," whatever your wretched " Wirklichkeit " may have to say to the matter.

And where, I should like to know, is a glimmer of a scintilla of a hint that the missionary was a dissenter ? I claim him for my dear National Church.—Ever yours,
T. H. HUXLEY.

The following is about a document which he had forgotten that he wrote :—

HODESLEA, EASTBOURNE,
Nov. 24, 1892.

MY DEAR DONNELLY—It is obvious that you have somebody in the Department who is an adept in the imitation of handwriting.

As there is no way of proving a negative, and I am too loyal to raise a scandal, I will just father the scrawl.

Positively, I had forgotten all about the business. I suppose because I did not hear who was appointed. It would be a good argument for turning people out of office after 65 ! But I have always had rather too much of the lawyer faculty of forgetting things when they are done with.

It was very jolly to have you here, and on principles of Christian benevolence you must not be so long in coming again.—Ever yours,   T. H. HUXLEY.

I do not remember being guilty of paying postage— but that doesn't count for much.

The following is an answer to one of the unexpected inquiries which would arrive from all quarters. A member of one of the religious orders working in the Church of England wrote for an authoritative statement on the following point, suggested by passages in section 5 of Chapter I. of the *Elementary Physiology* : — When the Blessed Sacrament, consisting, temporally and mundanely speaking, of a wheaten wafer and some wine, is received after about seven hours' fast, is it or is it not "voided like other meats"? In other words, does it not become completely absorbed for the sustenance of the body?

Huxley's help in this physiological question—and

his answer was to be used in polemical discussion—
was sought because an answer from him would be
decisive and would obviate the repetition of statements which to a Catholic were painfully irreverent.

HODESLEA, *Feb.* 3, 1892.

SIR—I regret that you have had to wait so long for a
reply to your letter of the 27th. Your question required
careful consideration, and I have been much occupied with
other matters.

You ask (1), whether the sacramental bread is or is not
" voided like other meats " ?

That depends on what you mean, firstly by " voided,"
and, secondly, by " other meats." Suppose any " meat "
(I take the word to include drink) to contain no
indigestible residuum, there need not be anything
" voided " at all—if by " voiding " is meant expulsion
from the lower intestine.

Such a meat might be " completely absorbed for the
sustenance of the body." Nevertheless, its elements, in
fresh combinations, would be eventually " voided " through
other channels, *e.g.* the lungs and kidneys. Thus I should
say that under normal circumstances all " meats " (that is
to say, the material substance of them) are voided sooner
or later.

Now, as to the particular case of the sacramental wafer
and wine. Taking their composition and the circumstances
of administration to be as you state them, it is my opinion
that a small residuum will be left undigested, and will be
voided by the intestine, while by far the greater part will
be absorbed and eventually " voided " by the lungs, skin,
and kidneys.

If any one asserts that the wafer and wine are voided
by the intestine as such, that the " pure flour and water "
of which the wafer consists pass out unchanged, I am of
opinion he is in error.

On the other hand, if any one maintains that the material substance of the wafer persists, while its accidents change, within the body, and that this identical substance is sooner or later voided, I do not see how he is to be driven out of that position by any scientific reasoning. On the contrary, there is every reason to believe that the elementary particles of the wafer and of the wine which enter the body never lose their identity, or even alter their mass. If one could see one of the atoms of carbon which enter into the composition of the wafer, I conceive it could be followed the whole way—from the mouth to the organ by which it escapes—just as a bit of floating charcoal might be followed into, through, and out of a whirlpool.

On October 6, 1892, died Lord Tennyson. In the course of his busy life, Huxley had not been thrown very closely into contact with him; they would meet at the Metaphysical Society, of which Tennyson was a silent member; and in the *Life of Tennyson* two occasions are recorded on which Huxley visited him :—

Nov. 11, 1871.—Mr. Huxley and Mr. Knowles arrived here (Aldworth) on a visit. Mr. Huxley was charming. We had much talk. He was chivalrous, wide, and earnest, so that one could not but enjoy talking with him. There was a discussion on George Eliot's humility. Huxley and A. both thought her a humble woman, despite a dogmatic manner of assertion that had come upon her latterly in her writings. (*Op. cit.* ii. 110.)

March 17, 1873.—Professor Tyndall and Mr. Huxley called. Mr. Huxley seemed to be universal in his interest, and to have keen enjoyment of life. He spoke of *In Memoriam.* (*Ibid.* ii. 143.)

With this may be compared one of Mr. Wilfrid
Ward's reminiscences (*Nineteenth Century*, August
1896).

"Huxley once spoke strongly of the insight into
scientific method shown in Tennyson's *In Memoriam*,
and pronounced it to be quite equal to that of the
greatest experts."

This view of Tennyson appears again in a letter to
Sir M. Foster, the Secretary of the Royal Society :—

Was not Tennyson a Fellow of the Royal Society ?
If so, should not the President and Council take some
notice of his death and delegate some one to the funeral
to represent them ? Very likely you have thought of it
already.

He was the only modern poet, in fact I think the only
poet since the time of Lucretius, who has taken the trouble
to understand the work and tendency of the men of
science.

But this was not the only side from which he
regarded poetry. He had a keen sense for beauty,
the artistic perfection of expression, whether in poetry,
prose, or conversation. Tennyson's talk he described
thus : "Doric beauty is its characteristic—perfect
simplicity, without any ornament or anything artificial." And again, to quote Mr. Wilfrid Ward's
reminiscences :—

Tennyson he considered the greatest English master of
melody except Spenser and Keats. I told him of
Tennyson's insensibility to music, and he replied that it
was curious that scientific men, as a rule, had more
appreciation of music than poets or men of letters. He

told me of one long talk he had had with Tennyson, and added that immortality was the one dogma to which Tennyson was passionately devoted.

Of Browning, Huxley said : " He really has music in him. Read his poem *The Thrush* and you will see it. Tennyson said to me," he added, " that Browning had plenty of music *in* him, but he could not get it *out*."

EASTBOURNE, Oct. 15, 1892.

MY DEAR TYNDALL—I think you will like to hear that the funeral yesterday lacked nothing to make it worthy of the dead or the living.

Bright sunshine streamed through the windows of the nave, while the choir was in half gloom, and as each shaft of light illuminated the flower-covered bier as it slowly travelled on, one thought of the bright succession of his works between the darkness before and the darkness after. I am glad to say that the Royal Society was represented by four of its chief officers, and nine of the commonalty, including myself. Tennyson has a right to that, as the first poet since Lucretius who has understood the drift of science.

We have heard nothing of you and your wife for ages. Ask her to give us news, good news I hope, of both.

My wife is better than she was, and joins with me in love.—Ever yours affectionately,   T. H. HUXLEY.

On his way home from the funeral in Westminster Abbey, Huxley passed the time in the train by shaping out some lines on the dead poet, the form of them suggested partly by some verses of his wife's, partly by Schiller's

> Gib diesen Todten mir heraus,
> Ich muss ihn wieder haben,[1]

---

[1] *Don Carlos*, Sc. ix.

which came back to his mind in the Abbey. The lines were published in the *Nineteenth Century* for November 1892. He declared that he deserved no credit for the verses ; they merely came to him in the train.

His own comparison of them with the sheaf of professed poets' odes which also appeared in the same magazine, comes in a letter to his wife, to whom he sent the poem as soon as it appeared in print.

> I know you want to see the poem, so I have cut it and the rest out of the *Nineteenth* just arrived, and sent it.
>
> If I were to pass judgment upon it in comparison with the others, I should say, that as to style it is hammered, and as to feeling human.
>
> They are castings of much prettier pattern and of mainly poetico-classical educated-class sentiment. I do not think there is a line of mine one of my old working-class audience would have boggled over. I would give a penny for John Burns' thoughts about it. (*N.B.*— Highly impartial and valuable criticism.)

He also wrote to Professor Romanes, who had been moved by this new departure to send him a volume of his own poems : —

> HODESLEA, *Nov.* 3, 1892.
>
> MY DEAR ROMANES—I must send you a line to thank you very much for your volume of poems. A swift glance shows me much that has my strong sympathy—notably " Pater loquitur," which I shall read to my wife as soon as I get her back. Against all troubles (and I have had my share) I weigh a wife-comrade " treu und fest " in all emergencies.

I have a great respect for the Nazarenism of Jesus—very little for later " Christianity." But the only religion that appeals to me is prophetic Judaism. Add to it something from the best Stoics and something from Spinoza and something from Goethe, and there is a religion for men. Some of these days I think I will make a cento out of the works of these people.

I find it hard enough to write decent prose and have usually stuck to that. The " Gib diesen Todten " I am hardly responsible for, as it did itself coming down here in the train after Tennyson's funeral. The notion came into my head in the Abbey.—Ever yours very faithfully,   T. H. HUXLEY.

This winter also Sir R. Owen died, and was buried at Ham on December 23. The grave ends all quarrels, and Huxley intended to be present at the funeral. But as he wrote to Dr. Foster on the 23rd :—

I had a hard morning's work at University College yesterday, and what with the meeting of the previous evening and that infernal fog, I felt so seedy that I made up my mind to go straight home and be quiet. . . .

There has been a bitter north-easter all day here, and if the like has prevailed at Ham I am glad I kept out of it, as I am by no means fit to cope with anything of that kind to-day. I do not think I was bound to offer myself up to the manes of the departed, however satisfactory that might have been to the poor old man. Peace be with him !

But the old-standing personal differences between the two made it difficult for him to decide what to do with regard to a meeting to raise some memorial to the great anatomist. He writes again to Sir M. Foster, January 8, 1893 :—

What am I to do about the meeting about Owen's statue on the 21st? I do not wish to pose either as a humbugging approver or as a sulky disapprover. The man did honest work, enough to deserve his statue, and that is all that concerns the public.

And on the 18th :—

I am inclined to think that I had better attend the meeting at all costs. But I do not see why I should speak unless I am called upon to do so.

I have no earthly objection to say all that I honestly can of good about Owen's work—and there is much to be said about some of it—on the contrary, I should be well pleased to do so.

But I have no reparation to make ; if the business were to come over again, I should do as I did. My opinion of the man's character is exactly what it was, and under the circumstances there is a sort of hypocrisy about volunteering anything, which goes against my grain.

The best position for me would be to be asked to second the resolution for the statue—then the proposer would have the field of personal fiction and butter-boat all to himself.

To Sir W. H. Flower

Dec. 28, 1892.

I think you are quite right in taking an active share in the movement for the memorial. When a man is dead and can do no more harm, one must do a sum in

$$\text{subtraction, deserts and if the } x\text{'s are not all minus} \quad \frac{\text{merits}}{x+x+x}$$

quantities, give him credit accordingly. But I think that in your appeal, for which the Committee will be responsible, it is this balance of solid scientific merit—a

good big one in Owen's case after all deductions—which
should be alone referred to.   If you follow the example
of *Vanity Fair* and call him "a simple-minded man, who
had he been otherwise, would long ago have adorned a
title," some of us may choke.

Gladstone, Samuel of Oxford, and Owen belong to a
very curious type of humanity, with many excellent and
even great qualities and one fatal defect—utter untrust-
worthiness.   Peace be with two of them, and may the
political death of the third be speedy and painless !—
With our united best wishes, ever yours very faithfully,

T. H. HUXLEY.

And on January 22, 1893, he writes of the
meeting :—

MY DEAR HOOKER— . . . What queer corners one gets
into if one only lives long enough !   The grim humour
of the situation when I was seconding the proposal for a
statue to Owen yesterday tickled me a good deal.   I do
not know how they will report me in the *Times,* but if
they do it properly I think you will see that I said no
word upon which I could not stand cross-examination.

I chose the office of seconder in order that I might
clearly define my position and stop the mouths of blasphemers—who would have ascribed silence or absence to
all sorts of bad motives.

Whatever the man might be, he did a lot of first-rate
work, and now that he can do no more mischief he has a
right to his wages for it.

If I only live another ten years I expect to be made
a saint of myself.   " Many a better man has been made
a saint of," as old Davie Hume said to his housekeeper
when they chalked up " St. David's Street " on his wall.

We have been jogging along pretty well, but wife has
been creaky, and I got done up in a brutal London fog
struggling with the worse fog of the New University.

I am very glad you like my poetical adventure.—Ever yours affectionately,   T. H. HUXLEY.

This speech had an unexpected sequel. Owen's grandson was so much struck by it that he wrote asking Huxley to undertake a critical account of his anatomical work for his biography,—another most unexpected turn of events. It is not often that a conspicuous opponent of a man's speculations is asked to pass judgment upon his entire work.[1]

At the end of the year an anonymous attack upon the administration of the Royal Society was the occasion for some characteristic words on the endurance of abuse to his old friend, M. Foster, then Secretary of the Royal Society.

Dec. 5, 1892.

MY DEAR FOSTER—The braying of my donkey prevented me from sending a word of sympathy about the noise made by yours. . . . Let not thine heart be vexed because of these sons of Belial. It is all sound and fury with nothing at the bottom of it, and will leave no trace a year hence. I have been abused a deal worse—without the least effect on my constitution or my comfort.

In fact, I am told that Harrison is abusing me just now like a pickpocket in the *Fortnightly*, and I only make the philosophical reflection, No wonder ! and doubt if the reading it is worth half a crown.—Ever yours affectionately,   T. H. HUXLEY.

The following letter to Mr. Clodd, thanking him for the new edition of Bates' *Naturalist on the Amazons*, helps to remove a reproach sometimes

[1] See p. 309.

brought against the Royal Society, in that it ignored
the claims of distinguished men of Science to membership of the Society :—

HODESLEA, EASTBOURNE,
Dec. 9, 1892.

My DEAR MR. CLODD—Many thanks for the new
edition of " Bates." I was reading the Life last night
with great interest ; some of the letters you have printed
are admirable.

Lyell is hit off to the life. I never read a more
penetrating character-sketch. Hooker's letter of advice is
as sage as might be expected from a man who practised
what he preached about as much as I have done. I shall
find material for chaff the next time my old friend and
I meet.

I think you are a little hard on the Trustees of the
British Museum, and especially on the Royal Society.
The former are hampered by the Treasury and the Civil
Service regulations. If a Bates turned up now I doubt if
one could appoint him, however much one wished it,
unless he would submit to some idiotic examination. As
to the Royal Society, I undertake to say that Bates
might have been elected fifteen years earlier if he had so
pleased. But the Council cannot elect a man unless he
is proposed, and I always understood that it was the *res
angusta* which stood in the way.

It is the same with ——. (Twenty years ago) the
Royal Society awarded him the Royal Medal, which is
about as broad an invitation to join us as we could well
give a man. In fact, I do not think he has behaved well
in quite ignoring it. Formerly there was a heavy
entrance fee as well as the annual subscription. But a
dozen or fifteen years ago the more pecunious Fellows
raised a large sum of money for the purpose of abolishing
this barrier. At present a man has to pay only £3 a

year and no entrance. I believe the publications of the
Society, which he gets, will sell for more.[1]

So you see it is not the fault of the Royal Society if
anybody who ought to be in keeps out on the score of
means.—Ever yours very faithfully,

T. H. HUXLEY.

---

[1] The 'Fee Reduction Fund," as it is now called, enables the
Society to relieve a Fellow from the payment even of his annual
fee, so that being F.R.S. costs him nothing.

CHAPTER XII

1893

THE year 1893 was, save for the death of three old friends, Andrew Clark, Jowett, and Tyndall, one of the most tranquil and peaceful in Huxley's whole life. He entered upon no direct controversy; he published no magazine articles; to the general misapprehension of the drift of his Romanes Lecture he only replied in the comprehensive form of Prolegomena to a reprint of the lecture. He began to publish his scattered essays in a uniform series, writing an introduction to each volume. While collecting his "Darwiniana" for the second volume, he wrote to Mr. Clodd :—

HODESLEA, EASTBOURNE,
Nov. 18, 1892.

I was looking through *Man's Place in Nature* the other day. I do not think there is a word I need delete, nor anything I need add except in confirmation and extension of the doctrine there laid down. That is great good fortune for a book thirty years old, and one that a very shrewd friend of mine implored me not to publish, as it would certainly ruin all my prospects. I said, like the French fox-hunter in *Punch*, "I shall try."

The shrewd friend in question was none other than Sir William Lawrence, whose own experiences after publishing his book *On Man*, "which now might be read in a Sunday school without surprising anybody," are alluded to in vol. i. p. 257.

He had the satisfaction of passing on his unfinished work upon *Spirula* to efficient hands for completion ; and in the way of new occupation, was thinking of some day "taking up the threads of late evolutionary speculation" in the theories of Weismann and others,[1] while actually planning out and reading for a series of "Working-Men's Lectures on the Bible," in which he should present to the unlearned the results of scientific study of the documents, and do for theology what he had done for zoology thirty years before.

The scheme drawn out in his note-book runs as follows :—

    I. The subject and the method of treating it.
    II. Physical conditions :—the place of Palestine in the Old World.
    III. The Rise of Israel :—Judges, Samuel, Kings as far as Jeroboam II.
    IV. The Fall of Israel.
    V. The Rise and Progress of Judaism.   Theocracy.
    VI. The Final Dispersion.
    VII. Prophetism.
    VIII. Nazarenism.
    IX. Christianity.
    X. Muhammedanism.
    XI. and XII. The Mythologies.

---

[1] See letter of September 28, to Romanes.

Although this scheme was never carried out, yet it was constantly before Huxley's mind during the two years left to him. If Death, who had come so near eight years before, would go on seeming to forget him, he meant to use these last days of his life in an effort to illuminate one more portion of the field of knowledge for the world at large.

As the physical strain of the Romanes Lecture and his liability to loss of voice warned him against any future attempt to deliver a course of lectures, he altered his design and prepared to put the substance of these Lectures to Working-Men into a Bible History for young people. And indeed, he had got so far with his preparation, that the latter heading was down in his list of work for the last year of his life, 1895. But nothing of it was ever written. Until the work was actually begun, even the framework upon which it was to be shaped remained in his mind, and the copious marks in his books of reference were the mere guide-posts to a strong memory, which retained not words and phrases, but salient facts and the knowledge of where to find them again.

I find only two occasions on which he wrote to the *Times* this year; one, when the crusade was begun to capture the Board Schools of London for sectarianism, and it was suggested that, when on the first School Board, he had approved of some such definite dogmatic teaching. This he set right at once in the following letter of April 28, with which may be compared the letter to Lord Farrer of November 6, 1894.

In a leading article of your issue to-day you state, with perfect accuracy, that I supported the arrangement respecting religious instruction agreed to by the London School Board in 1871, and hitherto undisturbed. But you go on to say that "the persons who framed the rule" intended it to include definite teaching of such theological dogmas as the Incarnation.

I cannot say what may have been in the minds of the framers of the rule ; but, assuredly, if I had dreamed that any such interpretation could fairly be put upon it, I should have opposed the arrangement to the best of my ability.

In fact, a year before the rule was framed I wrote an article in the *Contemporary Review*, entitled "The Board Schools—what they can do, and what they may do," in which I argued that the terms of the Education Act excluded such teaching as it is now proposed to include. And I supported my contention by the following citation from a speech delivered by Mr. Forster at the Birkbeck Institution in 1870 :—

" I have the fullest confidence that in the reading and explaining of the Bible, what the children will be taught will be the great truths of Christian life and conduct, which all of us desire they should know, and that no effort will be made to cram into their poor little minds, theological dogmas which their tender age prevents them from understanding."

The other was on a lighter, but equally perennial point of interest, being nothing less than the Sea Serpent. In the *Times* of January 11, he writes, that while there is no reason against a fifty-foot serpent existing as in Cretaceous seas, still the evidence for its existence is entirely inconclusive. He goes on to tell how a scientific friend's statement once almost convinced him until he read the quartermaster's deposition, which was supposed to corroborate it. The details made the circumstances alleged by the former impossible, and on pointing this out, he heard no more of the story, which was a good example of the mixing up of observations with conclusions drawn from them.

And on the following day he replies to another such detailed story—

Admiral Mellersh says, " I saw a huge snake, at least 18 feet long," and I have no doubt he believes he is simply stating a matter of fact.   Yet his assertion involves a hypothesis of the truth of which I venture to be exceedingly doubtful.   How does he know that what he saw was a snake?   The neighbourhood of a creature of this kind, within axe-stroke, is hardly conducive to calm scientific investigation, and I can answer for it that the discrimination of genuine sea-snakes in their native element from long-bodied fish is not always easy.   Further, that " back fin " troubles me ; looks, if I may say so, very fishy.

If the caution about mixing up observations with conclusions, which I ventured to give yesterday, were better attended to, I think we should hear very little either about antiquated sea-serpents or new " mesmerism."

It is perhaps not superfluous to point out that in this, as in other cases of the marvellous, he did not merely pooh-pooh a story on the ground of its antecedent improbability, but rested his acceptance or rejection of it upon the strength of the evidence adduced.   On the other hand, the weakness of such evidence as was brought forward time after time, was a justification for refusing to spend his time in listening to similar stories based on similar testimony.

Among the many journalistic absurdities which fall in the way of celebrities, two which happened this year are worth recording; the one on account of its intrinsic extravagance, which succeeded nevertheless in taking in quite a number of sober folk; the other on account of the letter it drew from Huxley about his cat. The former appeared in the shape of a highly-spiced advertisement about certain Manx Mannikins, which could walk, draw, play, in fact do everything but speak—were living pets which might be kept by any one, and indeed Professor Huxley was the possessor of a remarkably fine pair of them. Apply, enclosing stamps etc. Of course, the wonderful mannikins were nothing more than the pair of hands which anybody could dress up according to the instructions of the advertiser; but it was astonishing how many estimable persons took them for some *lusus naturæ*. A similar advertisement in 1880 had been equally successful, and one exalted personage wrote by the hand of a secretary to say what pleasure and interest had been excited by the description of these strange creatures, and begging Professor Huxley to state if the account was true. Accordingly on January 27 he writes to his wife, who was on a visit to her daughter :—

Yesterday two ladies called to know if they could see the Manx Mannikins. I think of having a board put up to say that in the absence of the Proprietress the show is closed.

The other incident was a request for any remarks which might be of use in an article upon the Home Pets of Celebrities. I give the letter written in answer to this, as well as descriptions of the same cat's goings-on in the absence of its mistress.

To Mr. J. G. Kitton

HODESLEA, *April* 12, 1893.

A long series of cats has reigned over my household for the last forty years, or thereabouts, but I am sorry to say that I have no pictorial or other record of their physical and moral excellences.

The present occupant of the throne is a large, young, grey Tabby—Oliver by name. Not that he is in any sense a protector, for I doubt whether he has the heart to kill a mouse. However, I saw him catch and eat the first butterfly of the season, and trust that this germ of courage, thus manifested, may develop with age into efficient mousing.

As to sagacity, I should say that his judgment respecting the warmest place and the softest cushion in a room is infallible—his punctuality at meal times is admirable ; and his pertinacity in jumping on people's shoulders, till they give him some of the best of what is going, indicates great firmness.

To his Youngest Daughter

HODESLEA EASTBOURNE,
Jan. 8, 1893.

I wish you would write seriously to M——. She is not behaving well to Oliver. I have seen handsomer kittens, but few more lively, and energetically destructive. Just now he scratched away at something that M—— says cost 13s. 6d. a yard—and reduced more or less of it to combings.

M—— therefore excludes him from the dining-room, and all those opportunities of higher education which he would naturally have in *my* house.

I have argued that it is as immoral to place 13s. 6d. a yard-nesses within reach of kittens as to hang bracelets and diamond rings in the front garden. But in vain. Oliver is banished—and the protector (not Oliver) is sat upon.—In truth and justice aid your Pa.

[This letter is embellished with fancy portraits of

Oliver when most quiescent (tail up ; ready for action).

O. as polisher (tearing at the table leg).

O. as plate basket investigator.

O. as gardener (destroying plants in a pot).

O. as stocking knitter (a wild tangle of cat and wool).

O. as political economist making good for trade at 13s. 6d. a yard (pulling at a hassock).]

The following to Sir John Evans refers to a piece of temporary forgetfulness.

HODESLEA, EASTBOURNE,
March 19, 1893.

MY DEAR EVANS—It is curious what a difference there is between intentions and acts, especially in the matter of sending cheques. The moment I saw the project of the Lawes and Gilbert testimonial in the *Times*, I sent my contribution in imagination—and it is only the arrival of this circular which has waked me up to the necessity of supplementing my ideal cheque by the real one inclosed. —Ever yours very faithfully,   T. H. HUXLEY.

Reference has been made to the writing of the Romanes Lecture in 1892. Mr. Gladstone had already consented to deliver the first lecture in that year ; and early in the summer Professor Romanes sounded Huxley to find out whether he would undertake the second lecture for 1893.   Huxley suggested a possible bar in his precarious health ; but subject to this possibility, if the Vice-Chancellor did not regard it as a complete disability, was willing to accept a formal invitation.

Professor Romanes reassured him upon this point, and further begged him, if possible, to be ready to step into the breach if Mr. Gladstone should be prevented from lecturing in the following autumn. The situation became irresistible, and the second of the following letters to Mr. Romanes displays no more hesitation.

To Professor Romanes

HODESLEA, *June* 3, 1892.

I should have written to you yesterday, but the book did not arrive till this morning.   Very many thanks for it.   It looks appetising, and I look forward to the next course.

As to the Oxford lecture, "Verily, thou almost persuadest me," though I thought I had finished lecturing. I really should like to do it ; but I have a scruple about accepting an engagement of this important kind, which I might not be able to fulfil.

I am astonishingly restored, and have not had a trace of heart trouble for months.   But I am quite aware that I am, physically speaking, on good behaviour — and maintain my condition only by taking an amount of care which is very distasteful to me.

Furthermore, my wife's health is, I am sorry to say, extremely precarious. She was very ill a fortnight ago, and to my very great regret, as well as hers, we are obliged to give up our intended visit to Balliol to-morrow She is quite unfit to travel, and I cannot leave her here alone for three days.

I think the state of affairs ought to be clear to the Vice-Chancellor. If, in his judgment, it constitutes no hindrance, and he does me the honour to send the invitation, I shall accept it.

To the Same

HODESLEA, *June* 7, 1892.

I am afraid that age hath not altogether cleared the spirit of mischief out of my blood ; and there is something so piquant in the notion of my acting as substitute for Gladstone that I will be ready if necessity arises.

Of course I will keep absolutely clear of Theology. But I have long had fermenting in my head, some notions about the relations of Ethics and Evolution (or rather the absence of such as are commonly supposed), which I think will be interesting to such an audience as I may expect. " Without prejudice," as the lawyers say, that is the sort of topic that occurs to me.

To the Same

HODESLEA, *Oct.* 30, 1892.

I had to go to London in the middle of last week about the Gresham University business, and I trust I have put a very long nail into the coffin of that scheme. For which good service you will forgive my delay in replying to your letter. I read all about your show—why not call it " George's Gorgeous," *tout court ?*

I should think that there is no living man, who, on such an occasion, could intend and contrive to say so much and so well (in form) without ever rising above the level of antiquarian gossip.

My lecture would have been ready if the G.O.M. had failed you, but I am very glad to have six months' respite, as I now shall be able to write and rewrite it to my heart's content.

I will follow the Gladstonian precedent touching cap and gown—but I trust the Vice-Chancellor will not ask me to take part in a "Church Parade" and read the lessons. I couldn't—really.

As to the financial part of the business, to tell you the honest truth, I would much rather not be paid at all for a piece of work of this kind. I am no more averse to turning an honest penny by my brains than any one else in the ordinary course of things—quite the contrary ; but this is not an ordinary occasion. However, this is a pure matter of taste, and I do not want to set a precedent which might be inconvenient to other people—so I agree to what you propose.

By the way, is there any type-writer who is to be trusted in Oxford ? Some time ago I sent a MS. to a London type-writer, and to my great disgust I shortly afterwards saw an announcement that I was engaged on the topic.

On the following day he writes to his wife, who was staying with her youngest daughter in town :—

The Vice-Chancellor has written to me and I have fixed May—exact day by and by. Mrs. Romanes has written a crispy little letter to remind us of our promise to go there, and I have chirrupped back.

The " chirrup " ran as follows :—

HODESLEA, *Nov.* 1, 1892.

MY DEAR MRS. ROMANES—I have just written to the Vice-Chancellor to say that I hope to be at his disposition any time next May.

My wife is "larking"—I am sorry to use such a word, but what she is pleased to tell me of her doings leaves me no alternative—in London, whither I go on Thursday to fetch her back—in chains, if necessary. But I know, in the matter of being "taken in and done for" by your hospitable selves, I may, for once, speak for her as well as myself.

Don't ask anybody above the rank of a younger son of a Peer—because I shall not be able to go in to dinner before him or her—and that part of my dignity is naturally what I prize most.

Would you not like me to come in my P.C. suit? All ablaze with gold, and costing a sum with which I could buy, oh! so many books!

Only if your late experiences should prompt you to instruct your other guests not to contradict me—don't. I rather like it.—Ever yours very faithfully,

T. H. HUXLEY.

Bon Voyage! You can tell Mr. Jones[1] that I will have him brought before the Privy Council and fined, as in the good old days, if he does not treat you properly.

This letter was afterwards published in Mrs. Romanes' Life of her husband, and three letters on that occasion, and particularly that in which Huxley tried to guard her from any malicious interpretation of his jests, are to be found on p. 332.

On the afternoon of May 18, 1893, he delivered

---

[1] The hotel-keeper in Madeira.

at Oxford his Romanes Lecture, on "Evolution and Ethics," a study of the relation of ethical and evolutionary theory in the history of philosophy, the text of which is that while morality is necessarily a part of the order of nature, still the ethical principle is opposed to the self-regarding principle on which cosmic evolution has taken place. Society is a part of nature, but would be dissolved by a return to the natural state of simple warfare among individuals. It follows that ethical systems based on the principles of cosmic evolution are not logically sound. A study of the essays of the foregoing ten years will show that he had more than once enunciated this thesis, and it had been one of the grounds of his long-standing criticism of Mr. Spencer's system.

The essence of this criticism is given in portions of two letters to Mr. F. J. Gould, who, when preparing a pamphlet on "Agnosticism writ Plain" in 1889, wrote to inquire what was the dividing line between the two Agnostic positions.

As between Mr. Spencer and myself, the question is not one of "a dividing line," but of entire and complete divergence as soon as we leave the foundations laid by Hume, Kant, and Hamilton, who are *my* philosophical forefathers. To my mind the "Absolute" philosophies were finally knocked on the head by Hamilton; and the "Unknowable" in Mr. Spencer's sense is merely the Absolute *redivivus*, a sort of ghost of an extinct philosophy, the name of a negation hocus-pocussed into a sham thing. If I am to talk about that of which I have no knowledge at all, I prefer the good old word *God*, about which there is no scientific pretence.

To my mind Agnosticism is simply the critical attitude of the thinking faculty, and the definition of it
should contain no dogmatic implications of any kind.
I, for my part, do not know whether the problem of
existence is insoluble or not ; or whether the ultimate
cause (if there be such a thing) is unknown or not. That
of which I am certain is, that no satisfactory solution of
this problem has been offered, and that, from the nature
of the intellectual faculty, I am unable to conceive that
such a solution will ever be found. But on that, as on
all other questions, my mind is open to consider any new
evidence that may be offered.

And later :—

I have long been aware of the manner in which my
views have been confounded with those of Mr. Spencer,
though no one was more fully aware of our divergence
than the latter. Perhaps I have done wrongly in letting
the thing slide so long, but I was anxious to avoid a
breach with an old friend. . . .

Whether the Unknowable or any other Noumenon
exists or does not exist, I am quite clear I have no
knowledge either way. So with respect to whether there
is anything behind Force or not, I am ignorant ; I neither
affirm nor deny. The tendency to idolatry in the human
mind is so strong that *faute de mieux* it falls down and
worships negative abstractions, as much the creation of
the mind as the stone idol of the hands. The one object
of the Agnostic (in the true sense) is to knock this tendency
on the head whenever or wherever it shows itself. Our
physical science is full of it.

Nevertheless, the doctrine seemed to take almost
everybody by surprise. The drift of the lecture was
equally misunderstood by critics of opposite camps.
Huxley was popularly supposed to hold the same

views as Mr. Spencer—for were they not both
Evolutionists? On general attention being called to
the existing difference between their views, some
jumped to the conclusion that Huxley was offering a
general recantation of evolution, others that he had
discarded his former theories of ethics. On the one
hand he was branded as a deserter from free thought;
on the other, hailed almost as a convert to orthodoxy.
It was irritating, but little more than he had expected.
The conditions of the lecture forbade any reference to
politics or religion; hence much had to be left unsaid,
which was supplied next year in the Prolegomena
prefacing the re-issue of the lecture.

After all possible trimming and compression, he
still feared the lecture would be too long, and would
take more than an hour to deliver, especially if the
audience was likely to be large, for the numbers must
be considered in reference to the speed of speaking.
But he had taken even more pains than usual with it.
"The Lecture," he writes to Professor Romanes on
April 19, "has been in type for weeks, if not months,
as I have been taking an immensity of trouble over
it. And I can judge of nothing till it is in type."
But this very precaution led to unexpected com-
plications. When the proposition to lecture was first
made to him, he was not sent a copy of the statute
ordering that publication in the first instance should
lie with the University Press; and in view of the
proviso that "the Lecturer is free to publish on his
own behalf in any other form he may like," he had

taken Prof. Romanes' original reference to publication by the Press to be a subsidiary request to which he gladly assented. However, a satisfactory arrangement was speedily arrived at with the publishers; Huxley remarking :—

All I have to say is, do not let the University be in any way a loser by the change. If the V-C. thinks there is any risk of this, I will gladly add to what Macmillan pays. That matter can be settled between us.

However, he had not forgotten the limitation of his subject in respect of religion and politics, and he repeatedly refers to his careful avoidance of these topics as an "egg-dance." And wishing to reassure Mr. Romanes on this head, he writes on April 22 :—

There is no allusion to politics in my lecture, nor to any religion except Buddhism, and only to the speculative and ethical side of that. If people apply anything I say about these matters to modern philosophies, except evolutionary speculation, and religions, that is not my affair. To be honest, however, unless I thought they would, I should never have taken all the pains I have bestowed on these 36 pages.

But these words conjured up terrible possibilities, and Mr. Romanes wrote back in great alarm to ask the exact state of the case. The two following letters show that the alarm was groundless :—

HODESLEA, *April* 26, 1893.

MY DEAR ROMANES—I fear, or rather hope, that I have given you a very unnecessary scare.

You may be quite sure, I think, that, while I should have refused to give the lecture if any pledge of a special

character had been proposed to me, I have felt very
strongly bound to you to take the utmost care that no
shadow of a just cause for offence should be given, even to
the most orthodox of Dons.

It seems to me that the best thing I can do is to send
you the lecture as it stands, notes and all. But please
return it within two days at furthest, and consider it
strictly confidential between us two (I am not excluding
Mrs. Romanes, if she cares to look at the paper). No
consideration would induce me to give any ground for the
notion that I had submitted the lecture to any one but
yourself.

If there is any phrase in the lecture which you think
likely to get you into trouble, out it shall come or be
modified in form.

If the whole thing is too much for the Dons' nerves—
I am no judge of their delicacy—I am quite ready to give
up the lecture.

In fact I do not know whether I shall be able to make
myself heard three weeks hence, as the influenza has left its
mark in hoarseness and pain in the throat after speaking.

So you see if the thing is altogether too wicked there
is an easy way out of it.—Ever yours very faithfully,

T. H. HUXLEY.

HODESLEA, *April* 28, 1893.

MY DEAR ROMANES—My mind is made easy by such a
handsome acquittal from you and the Lady Abbess, your
coadjutor in the Holy Office.

My wife, who is my inquisitor and confessor in ordinary,
has gone over the lecture twice, without scenting a heresy,
and if she and Mrs. Romanes fail—a fico for a mere male
don's nose !

From the point of view of the complete argument, I
agree with you about note 19. But the dangers of open
collision with orthodoxy on the one hand and Spencer on
the other. increased with the square of the enlargement

of the final pages, and I was most anxious for giving no handle to any one who might like to say I had used the lecture for purposes of attack. Moreover, in spite of all reduction, the lecture is too long already.

But I think it not improbable that in spite of my meekness and peacefulness, neither the one side nor the other will let me alone. And then you see, I shall have an opportunity of making things plain, under no restriction. You will not be responsible for anything said in the second edition, nor can the Donniest of Dons grumble. —Ever yours very faithfully,   T. H. HUXLEY.

The double negative is Shakspearian. See *Hamlet*, act ii. sc. 2.

Unfortunately for the entire success of the lecture, he was suffering from the results of influenza, more especially a loss of voice. He writes (April 18) :—

After getting through the winter successfully I have had the ill-fortune to be seized with influenza. I believe I must have got it from the microbes haunting some of the three hundred doctors at the Virchow dinner.[1]

I had next to no symptoms except debility, and though I am much better I cannot quite shake that off. As usual with me it affects my voice. I hope this will get right before this day month, but I expect I shall have to nurse it. I do not want to interfere with any of your hospitable plans, and I think if you will ensure me quiet on the morning of the 18th (I understand the lecture is in the afternoon) it will suffice. After the thing is over I am ready for anything from pitch and toss onwards.

Two more letters dated before the 18th of May touch on the circumstances of the lecture. One is to his son-in-law, John Collier; the other to his old

---

[1] On the 16th March.

friend Tyndall, the last he ever wrote him, and containing a cheery reference to the advance of old age.

HODESLEA, EASTBOURNE,
May 9, 1893.

MY DEAR JACK—. . . M—— is better, and I am getting my voice back. But may St. Ernulphus' curse descend on influenza microbes! They tried to work their way out at my nose, and converted me into a disreputable Captain Costigan-looking person ten days ago. Now they are working at my lips.

For the credit of the family I hope I shall be more reputable by the 18th.

I hope you will appreciate my dexterity. The lecture is a regular egg-dance. That I should discourse on Ethics to the University of Oxford and say all I want to say, without a word anybody can quarrel with, is decidedly the most piquant occurrence in my career. . . .—Ever yours affectionately,   PATER.

TO PROFESSOR TYNDALL

P.S. to be read first.

EASTBOURNE, *May* 15, 1893.

MY DEAR TYNDALL—There are not many apples (and those mostly of the crab sort) left upon the old tree, but I send you the product of the last shaking. Please keep it out of any hands but your wife's and yours till Thursday, when I am to "stand and deliver" it, if I have voice enough, which is doubtful. The sequelæ of influenza in my case have been mostly pimples and procrastination, the former largely on my nose, so that I have been a spectacle. Besides these, loss of voice. The pimples are mostly gone and the procrastination is not much above normal, but what will happen when I try to fill the Sheldonian Theatre is very doubtful.

Who would have thought thirty-three years ago, when the great " Sammy " fight came off, that the next time I should speak at Oxford would be in succession to Gladstone, on "Evolution and Ethics" as an invited lecturer ?

There was something so quaint about the affair that I really could not resist, though the wisdom of putting so much strain on my creaky timbers is very questionable. Mind you wish me well through it at 2.30 on Thursday.

I wish we could have better news of you. As to dying by inches, that is what we are all doing, my dear old fellow ; the only thing is to establish a proper ratio between inch and time. Eight years ago I had good reason to say the same thing of myself, but my inch has lengthened out in a most extraordinary way. Still I confess we are getting older ; and my dear wife has been greatly shaken by repeated attacks of violent pain which seizes her quite unexpectedly. I am always glad, both on her account and my own, to get back into the quiet and good air here as fast as possible, and in another year or two, if I live so long, I shall clear out of all engagements that take me away. . . .   T. H. HUXLEY.

Not to be answered, and you had better get Mrs. Tyndall to read it to you or you will say naughty words about the scrawl.

Sanguine as he had resolved to be about the recovery of his voice, his fear lest "1000 out of the 2000 won't hear" was very near realisation. The Sheldonian Theatre was thronged before he appeared on the platform, a striking presence in his D.C.L. robes, and looking very leonine with his silvery gray hair sweeping back in one long wave from his forehead, and the rugged squareness of his features tempered by the benignity of an old age which has

seen much and overcome much.   He read the lecture
from a printed copy, not venturing, as he would have
liked, upon the severe task of speaking it from
memory, considering its length and the importance
of preserving the exact wording.   He began in a
somewhat low tone, nursing his voice for the second
half of the discourse.   From the more distant parts
of the theatre came several cries of "speak up"; and
after a time a rather disturbing migration of eager
undergraduates began from the galleries to the body
of the hall.   The latter part was indeed more audible
than the first; still a number of the audience were
disappointed in hearing imperfectly.   However, the
lecture had a large sale; the first edition of 2000 was
exhausted by the end of the month; and another 700
in the next ten days.

After leaving Oxford, and paying a pleasant visit
to one of the Fannings (his wife's nephew) at Tew,
Huxley intended to visit another of the family, Mrs.
Crowder, in Lincolnshire, but on reaching London
found himself dead beat, and had to retire to Eastbourne, whence he writes to Sir M. Foster and to
Mr. Romanes.

HODESLEA, *May* 26, 1893.

MY DEAR FOSTER—Your letter has been following me
about.   I had not got rid of my influenza at Oxford, so
the exertion and the dinner parties together played the
deuce with me.

We had got so far as the Great Northern Hotel on
our way to some connections in Lincolnshire, when I had
to give it up and retreat here to begin convalescing again.

I do not feel sure of coming to the Harvey affair after all. But if I do, it will be alone, and I think I had better accept the hospitality of the college ; which will by no means be so jolly as Shelford, but probably more prudent, considering the necessity of dining out.

The fact is, my dear friend, I am getting old.

I am very sorry to hear you have been doing your influenza also. It's a beastly thing, as I have it, no symptoms except going flop.—Ever yours,

T. H. HUXLEY.

Nobody sees that the lecture is a very orthodox production on the text (if there is such a one), "Satan the Prince of this world."

I think the remnant of influenza microbes must have held a meeting in my *corpus* after the lecture, and resolved to reconquer the territory. But I mean to beat the brutes.

" I shall be interested," he writes to Mr. Romanes, "in the article on the lecture. The papers have been asinine." This was an article which Mr. Romanes had told him was about to appear in the *Oxford Magazine.* And on the 30th he writes again :—

Many thanks for the *Oxford Magazine.* The writer of the article is about the only critic I have met with yet who understands my drift. My wife says it is a " sensible " article, but her classification is a very simple one—sensible articles are those that contain praise, " stupid " those that show insensibility to my merits !

Really I thought it very sensible, without regard to the plums in the pudding.

But the criticism, " sensible " not merely in the humorous sense, which he most fully appreciated was that of Professor Seth, in a lecture entitled "Man and Nature." He wrote to him on October 27 :—

DEAR PROFESSOR SETH—A report of your lecture on "Man and Nature" has just reached me. Accept my cordial thanks for defending me, and still more for understanding me.

I really have been unable to understand what my critics have been dreaming of when they raise the objection that the ethical process being part of the cosmic process cannot be opposed to it.

They might as well say that artifice does not oppose nature, because it is part of nature in the broadest sense.

However, it is one of the conditions of the "Romanes Lecture" that no allusion shall be made to religion or politics. I had to make my omelette without breaking any of those eggs, and the task was not easy.

The prince of scientific expositors, Faraday, was once asked, "How much may a popular lecturer suppose his audience knows?" He replied emphatically, "*Nothing.*" Mine was not exactly a popular audience, but I ought not to have forgotten Faraday's rule.—Yours very faithfully,

T. H. HUXLEY.

A letter of congratulation to Lord Farrer on his elevation to the peerage contains an ironical reference to the general tone of the criticisms on his lecture :—

HODESLEA, *June* 5, 1893.

CI DEVANT CITOYEN PÉTION (autrefois le vertueux)— You have lost all chance of leading the forces of the County Council to the attack of the Horse-Guards.

You will become an émigré, and John Burns will have to content himself with the heads of the likes of me. As the Jacobins said of Lavoisier, the Republic has no need of men of science.

But this prospect need not interfere with sending our hearty congratulations to Lady Farrer and yourself.

As for your criticisms, don't you know that I am become a reactionary and secret friend of the clerics ?

My lecture is really an effort to put the Christian doctrine that Satan is the Prince of this world upon a scientific foundation.

Just consider it in this light, and you will understand why I was so warmly welcomed in Oxford. (N.B.—The only time I spoke before was in 1860, when the great row with Samuel came off ! !)—Ever yours very faithfully,

T. H. HUXLEY.

HODESLEA, EASTBOURNE,
July 15, 1893.

MY DEAR SKELTON—I fear I must admit that even a Gladstonian paper occasionally tells the truth. They never mean to, but we all have our lapses from the rule of life we have laid down for ourselves, and must be charitable.

The fact is, I got influenza in the spring, and have never managed to shake right again, any tendency that way being well counteracted by the Romanes lecture and its accompaniments.

So we are off to the Maloja to-morrow. It mended up the shaky old heart-pump five years ago, and I hope will again.

I have been in Orkney, and believe in the air, but I cannot say quite so much for the scenery. I thought it just a wee little bit, shall I say, bare ? But then I have a passion for mountains.

I shall be right glad to know what your H.O.M. [1] has to say about Ethics and Evolution. You must remember that my lecture was a kind of egg-dance. Good manners bound me over to say nothing offensive to

---

[1] The "Old Man of Hoy," a pseudonym under which Sir J. Skelton wrote.

the Christians in the amphitheatre (I was in the arena), and truthfulness, on the other hand, bound me to say nothing that I did not fully mean.   Under these circumstances one has to leave a great many i's undotted and t's uncrossed.

Pray remember me very kindly to Mrs. Skelton, and believe me—Yours ever,   T. H. HUXLEY.

And again on Oct. 17 :—

Ask your Old Man of Hoy to be so good as to suspend judgment until the Lecture appears again with an appendix in that collection of volumes the bulk of which appals me.

Didn't I see somewhere that you had been made Poor Law pope, or something of the sort ?   I congratulate the poor more than I do you, for it must be a weary business trying to mend the irremediable.   (No, I am *not* glancing at the whitewashing of Mary.)

Here may be added two later letters bearing in part upon the same subject :—

HODESLEA, EASTBOURNE,
March 23, 1894.

DEAR SIR—I ought to have thanked you before now for your letter about Nietzsche's works, but I have not much working time, and I find letter-writing a burden, which I am always trying to shirk.

I will look up Nietzsche, though I must confess that the profit I obtain from German authors on speculative questions is not usually great.

As men of research in positive science they are magnificently laborious and accurate.   But most of them have no notion of style, and seem to compose their books with a pitchfork.

There are two very different questions which people fail to discriminate.   One is whether evolution accounts for morality, the other whether the principle of evolution in general can be adopted as an ethical principle.

The first, of course, I advocate, and have constantly insisted upon. The second I deny, and reject all so-called evolutional ethics based upon it.—I am yours faithfully,

T. H. HUXLEY.

Thomas Common, Esq.

HODESLEA, *August* 31, 1894.

DEAR PROFESSOR SETH—I have come to a stop in the issue of my essays for the present, and I venture to ask your acceptance of the set which I have desired my publishers to send you.

I hope that at present you are away somewhere, reading novels or otherwise idling, in whatever may be your pet fashion.

But some day I want you to read the " Prolegomena " to the reprinted Romanes Lecture.

Lately I have been re-reading Spinoza (much read and little understood in my youth).

But that noblest of Jews must have planted no end of germs in my brains, for I see that what I have to say is in principle what he had to say, in modern language.— Ever yours very faithfully,   T. H. HUXLEY.

The following letters with reference to the long unfinished memoir on " Spirula " for the *Challenger* reports tell their own story. Huxley was very glad to find some competent person to finish the work which his illness had incapacitated him from completing himself. It had been a burden on his conscience ; and now he gladly put all his plates and experience at the disposal of Professor Pelseneer, though he had nothing written and would not write anything. He had no wish to claim even joint authorship for the completed paper ; when the question was first raised, he desired merely that it should be stated that such and such drawings were made by him ; but when Professor Pelseneer insisted that both names should appear as joint authors, he consented to this solution of the question.

HODESLEA, *Sept.* 17, 1893.

DEAR MR. MURRAY [1]—If the plates of *Spirula* could be turned to account a great burthen would be taken off my mind.

Professor Pelseneer is every way competent to do justice to the subject ; and he has just what I needed, namely another specimen to check and complete the work ; and besides that, the physical capacity for dissection and close observation, of which I have had nothing left since my long illness.

Will you be so good as to tell Professor Pelseneer that I shall be glad to place the plates at his disposal and to give him all the explanations I can of the drawings, whenever it may suit his convenience to take up the work ?

Nothing beyond mere fragments remained of the specimen.—I am, yours very faithfully,

T. H. HUXLEY.

I return Pelseneer's letter.

HODESLEA, *Sept.* 30, 1893.

DEAR PROFESSOR PELSENEER—I send herewith (by this post) a full explanation of the plates of *Spirula* (including those of which you have unlettered copies). I trust you will not be too much embarrassed by my bad

---

[1] Now K.C.B. ; Director of the " Reports of the *Challenger*."

handwriting, which is a plague to myself as well as to other people.

My hope is that you will be good enough to consider these figures as materials placed in your hands, to be made useful in the memoir on *Spirula*, which I trust you will draw up, supplying the defects of my work and checking its accuracy.

You will observe that a great deal remains to be done. The muscular system is untouched; the structure and nature of the terminal circumvallate papilla have to be made out; the lingual teeth must be re-examined; and the characters of the male determined. If I recollect rightly, Owen published something about the last point.

If I can be of any service to you in any questions that arise, I shall be very glad; but as I am putting the trouble of the work on your shoulders, I wish you to have the credit of it.

So far as I am concerned, all that is needful is to say that such and such drawings were made by me.—Ever yours very faithfully, T. H. HUXLEY.

HODESLEA, *Oct.* 12, 1893.

DEAR PROFESSOR PELSENEER—I am very glad to hear from you that the homology of the cephalopod arms with the gasteropod foot is now generally admitted. When I advocated that opinion in my memoir on the "Morphology of the Cephalous Mollusca," some forty years ago, it was thought a great heresy.

As to publication; I am quite willing to agree to whatever arrangement you think desirable, so long as you are kind enough to take all trouble (but that of "consulting physician") off my shoulders. Perhaps putting both names to the memoir, as you suggest, will be the best way. I cannot undertake to write anything, but if you think I can be of any use as an adviser or critic, do not hesitate to demand my services. — Ever yours very faithfully, T. H. HUXLEY.

Although in February he had stayed several days in town with the Donnellys, who "take as much care of me as if I were a piece of old china," and had attended a levée and a meeting of his London University Association, had listened with interest to a lecture of Professor Dewar, who "made liquid oxygen by the pint," and dined at Marlborough House, the influenza had prevented him during the spring from fulfilling several engagements in London ; but after his return from Oxford he began to recruit in the fine weather, and found delightful occupation in putting up a rockery in the garden for his pet Alpine plants.

In mid June he writes to his wife, then on a visit to one of her daughters :—

What a little goose you are to go having bad dreams about me—who am like a stalled ox—browsing in idle comfort—in fact, idle is no word for it.   Sloth is the right epithet.   I can't get myself to do anything but potter in the garden, which is looking lovely.

On June 21 he went to Cambridge for the Harvey Celebration at Gonville and Caius College, and made a short speech.

The dinner last night (he writes) was a long affair, and I was the last speaker ; but I got through my speech very well, and was heard by everybody, I am told.

But as is the way with influenza, it was thrown off in the summer only to return the next winter, and on the eve of the Royal Society Anniversary Dinner ne writes to Sir M. Foster :—

I am in rather a shaky and voiceless condition, and
unless I am more up to the mark to-morrow morning I
shall have to forgo the dinner, and, what is worse, the
chat with you afterwards.

One consequence of the spring attack of influenza
was that this year he went once more to the Maloja,
staying there from July 21 to August 25.

> HODESLEA, EASTBOURNE,
> *July* 9, 1893.

MY DEAR HOOKER—What has happened to the *x*
meeting you proposed? However, it does not matter
much to me now, as Hames, who gave me a thorough
overhauling in London, has packed me off to the Maloja
again, and we start, if we can, on the 17th.

It is a great nuisance, but the dregs of influenza and
the hot weather between them have brought the weakness
of my heart to the front, and I am gravitating to the
condition in which I was five or six years ago. So I
must try the remedy which was so effectual last time.

We are neither of us very fit, and shall have to be
taken charge of by a courier. Fancy coming to that!

Let me be a warning to you, my dear old man. Don't
go giving lectures at Oxford and making speeches at
Cambridge, and above all things don't, oh don't go getting
influenza, the microbes of which would be seen under a
strong enough microscope to have this form.

[Sketch of an active little black demon.]

> T. H. HUXLEY.

Though not so strikingly as before, the high Alpine
air was again a wonderful tonic to him. His diary
still contains a note of occasional long walks; and
once more he was the centre of a circle of friends,
whose cordial recollections of their pleasant intercourse

afterwards found expression in a lasting memorial. Beside one of his favourite walks, a narrow pathway skirting the blue lakelet of Sils, was placed a gray block of granite. The face of this was roughly smoothed, and upon it was cut the following inscription :—

In memory of the illustrious English Writer and Naturalist, Thomas Henry Huxley, who spent many summers at the Kursaal, Maloja.

In a letter to Sir J. Hooker, of October 1, he describes the effects of his trip, and his own surprise at being asked to write a critical account of Owen's work :—

HODESLEA, EASTBOURNE,
Oct. 1, 1893.

MY DEAR HOOKER—I am no better than a Gadarene swine for not writing to you from the Maloja, but I was too procrastinatingly lazy to expend even that amount of energy. I found I could walk as well as ever, but unless I was walking I was everlastingly seedy, and the wife was unwell almost all the time. I am inclined to think that it is coming home which is the most beneficial part of going abroad, for I am remarkably well now, and my wife is very much better.

I trust the impaled and injudicious Richard [1] is none the worse. It is wonderful what boys go through (also what goes through them).

You will get all the volumes of my screeds. I was horrified to find what a lot of stuff there was—but don't acknowledge them unless the spirit moves you. . . . I

---

[1] Sir J. Hooker's youngest son, who had managed to spike himself on a fence.

think that on Natural Inequality of Man will be to your taste.

Three, or thirty, guesses and you shall not guess what I am about to tell you.

Rev. Richard Owen has written to me to ask me to write a concluding chapter for the biography of his grandfather—containing a "critical" estimate of him and his work ! ! ! Says he is moved thereto by my speech at the meeting for a memorial.

There seemed nothing for me to do but to accept as far as the scientific work goes. I declined any personal estimate on the ground that we had met in private society half a dozen times.

If you don't mind being bothered I should like to send you what I write and have your opinion about it.

You see Jowett is going or gone. I am very sorry we were obliged to give up our annual visit to him this year. But I was quite unable to stand the exertion, even if Hames had not packed me off. How one's old friends are dropping !

Romanes gave me a pitiable account of himself in a letter the other day. He has had an attack of hemiplegic paralysis, and tells me he is a mere wreck. That means that the worst anticipations of his case are being verified. It is lamentable.

Take care of yourself, my dear old friend, and with our love to you both, believe me, ever yours,

T. H. HUXLEY.

Not long after his return he received a letter from a certain G—— S——, who wrote from Southampton detailing a number of observations he had made upon the organisms to be seen with a magnifying glass in an infusion of vegetable matter, and as "an ignoramus," apologised for any appearance of conceit in so doing, while asking his advice as to the best means of improving his scientific knowledge. Huxley was much struck by the tone of the letter and the description of the experiments, and he wrote back :—

HODESLEA, *Nov.* 9, 1893.

SIR—We are all "ignoramuses" more or less—and cannot reproach one another.   If there were any sign of conceit in your letter, you would not get this reply.

On the contrary, it pleases me.   Your observations are quite accurate and clearly described—and to be accurate in observation and clear in description is the first step towards good scientific work.

You are seeing just what the first workers with the microscope saw a couple of centuries ago.

Get some such book as Carpenter's "On the Microscope" and you will see what it all means.

Are there no science classes in Southampton ?   There used to be, and I suppose is, a Hartley Institute.

If you want to consult books you cannot otherwise obtain, take this to the librarian, give him my compliments, and say I should be very much obliged if he would help you.—I am, yours very faithfully,

T. H. HUXLEY.

Great was Huxley's astonishment when he learned in reply that his correspondent was a casual dock labourer, and had but scanty hours of leisure in which to read and think and seek into the recesses of nature, while his means of observation consisted of a toy microscope bought for a shilling at a fair. Casting about for some means of lending the man a helping hand, he bethought him of the Science and Art Department, and wrote on December 30 to Sir J. Donnelly :—

The Department has feelers all over England—has it any at Southampton? And if it has, could it find out something about the writer of the letters I enclose? For a "casual docker" they are remarkable; and I think when you have read them you will not mind my bothering you with them. (I really have had the grace to hesitate.)

I have been puzzled what to do for the man. It is so much easier to do harm than good by meddling—and yet I don't like to leave him to "casual docking."

In that first letter he has got—on his own hook—about as far as Buffon and Needham 150 years ago.

And later to Professor Howes :—

HODESLEA, EASTBOURNE,
Feb. 12, 1894.

MY DEAR HOWES—Best thanks for unearthing the volumes of Milne-Edwards. I was afraid my set was spoiled.

I shall be still more obliged to you if you can hear of something for S——. There is a right good parson in his neighbourhood, and from what he tells me about S—— I am confirmed in my opinion that he is a very exceptional man, who ought to be at something better than porter's work for twelve hours a day.

The mischief is that one never knows how transplanting a tree, much less a man, will answer. Playing Providence is a game at which one is very apt to burn one's fingers.

However, I am going to try, and hope at any rate to do no harm to the man I want to help.—Ever yours very faithfully,　　　T. H. HUXLEY.

He was eventually offered more congenial occupation at the Natural History Museum in South

Kensington, but preferred not to enter into the
bonds of an unaccustomed office.

Meanwhile, through Sir John Donnelly, Huxley
was placed in communication with the Rev. Montague
Powell, who, at his request, called upon the docker;
and finding him a man who had read and thought
to an astonishing extent upon scientific problems,
and had a considerable acquaintance with English
literature, soon took more than a vicarious interest
in him.   Mr. Powell, who kept Huxley informed of
his talks and correspondence with G. S., gives a full
account of the circumstances in a letter to the
Spectator of July 13, 1895, from which I quote the
following words :—

> The Professor's object in writing was to ask me how
> best such a man could be helped, I being at his special
> request the intermediary.   So I suggested in the mean-
> while a microscope and a few scientific books.   In the
> course of a few days I received a splendid achromatic
> compound microscope and some books, which I duly
> handed over to my friend, telling him it was from an
> unknown hand.   " Ah," he said, " I know who that must
> be ; it can be no other than the greatest of living scientists ;
> it is just like him to help a tyro."

One small incident of this affair is perhaps worth
preserving as an example of Huxley's love of a
bantering repartee.   In the midst of the correspondence Mr. Powell seems suddenly to have been seized
by an uneasy recollection that Huxley had lately
received some honour or title, so he next addressed
him as " My dear Sir Thomas."   The latter, not to

be outdone, promptly replied with "My dear Lord Bishop of the Solent."

About the same time comes a letter to Mr. Knowles, based upon a paragraph from the gossiping column of some newspaper which had come into Huxley's hands :—

HODESLEA, EASTBOURNE,
Nov. 9, 1893.

Gossip of the Town.

"Professor Huxley receives 200 guineas for each of his articles for the *Nineteenth Century.*"

MY DEAR KNOWLES—I have always been satisfied with the *Nineteenth Century* in the capacity of paymaster, but I did not know how much reason I had for my satisfaction till I read the above !

Totting up the number of articles and multiplying by 200 it strikes me I shall be behaving very handsomely if I take £2000 for the balance due.

So sit down quickly, take thy cheque-book, and write five score, and let me have it at breakfast time to-morrow. I once got a cheque for £1000 at breakfast, and it ruined me morally.   I have always been looking out for another.

I hope you are all flourishing.   We are the better for Maloja, but more dependent on change of weather and other trifles than could be wished.   Yet I find myself outlasting those who started in life along with me.   Poor Andrew Clark and I were at Haslar together in 1846, and he was the younger by a year and a half.—Ever yours very faithfully,   T. H. HUXLEY.

All my time is spent in the co-ordination of my eruptions when I am an active volcano.

I hope you got the volumes which I told Macmillan to send you.

The following letter to Professor Romanes, whose failing eyesight was a premonitory symptom of the disease which proved fatal the next year, reads, so to say, as a solemn prelude to the death of three old friends this autumn—of Andrew Clark, his old comrade at Haslar, and cheery physician for many years; of Benjamin Jowett, Master of Balliol, whose acquaintance he had first made in 1851 at the Stanleys' at Harrow, and with whom he kept up an intimacy to the end of his life, visiting Balliol once or twice every year ; and, heaviest blow, of John Tyndall, the friend and comrade whose genial warmth of spirit made him almost claim a brother's place in early struggles and later success, and whose sudden death was all the more poignant for the cruel touch of tragedy in the manner of it :—

HODESLEA, *Sept.* 28, 1893.

MY DEAR ROMANES—We are very much grieved to hear such a bad account of your health. Would that we could achieve something more to the purpose than assuring you and Mrs. Romanes of our hearty sympathy with you both in your troubles. I assure you, you are much in our thoughts, which are·sad enough with the news of Jowett's, I fear, fatal attack.

I am almost ashamed to be well and tolerably active when young and old friends are being thus prostrated.

However, you have youth on your side, so do not give up, and wearisome as doing nothing may be, persist in it as the best of medicines.

At my time of life one should be always ready to stand at attention when the order to march comes ; but for the rest I think it well to go on doing what I can, as if F. M.

General Death had forgotten me. That must account for my seeming presumption in thinking I may some day "take up the threads" of late evolutionary speculation.— Ever yours very faithfully,   T. H. HUXLEY.

My wife joins with me in love and kind wishes to you both.

At the request of his friends, Huxley wrote for the *Nineteenth Century* a brief appreciation of his old comrade Tyndall—the tribute of a friend to a friend —and, difficult task though it was, touched on the closing scene, if only from a chivalrous desire to do justice to the long devotion which accident had so cruelly wronged :—

I am comforted (he writes to Sir J. Hooker on January 3) by your liking the Tyndall article. You are quite right, I shivered over the episode of the "last words," but it struck me as the best way of getting justice done to her, so I took a header. I am glad to see by the newspaper comments that it does not seem to have shocked other people's sense of decency.

The funeral took place on Saturday, December 9. There was no storm nor fog to make the graveside perilous for the survivors. In the Haslemere church-yard the winter sun shone its brightest, and the moorland air was crisp with an almost Alpine freshness as this lover of the mountains was carried to his last resting-place. But though he took no outward harm from that bright still morning, Huxley was greatly shaken by the event : "I was very much used up," he writes to Sir M. Foster on his return home two

days later, "to my shame be it said, far more than
my wife"; and on December 30 to Sir John
Donnelly :—

> Your kind letter deserved better than to have been left
> all this time without response, but the fact is, I came to
> grief the day after Christmas Day (no, we did *not* indulge
> in too much champagne). Lost my voice, and collapsed
> generally, without any particular reason, so I went to bed
> and stayed there as long as I could stand it, and now I
> am picking up again. The fact is, I suppose I had been
> running up a little account over poor old Tyndall. One
> does not stand that sort of wear and tear so well as one
> gets ancient.

On the same day he writes to Sir J. D. Hooker :—

> HODESLEA, EASTBOURNE,
> *Dec.* 30, 1893.
>
> MY DEAR HOOKER—You gave the geographers some
> uncommonly sane advice. I observe that the words about
> the "stupendous ice-clad mountains" you saw were hardly
> out of your mouth when —— coolly asserts that the
> Antarctic continent is a table-land ! "comparatively level
> country." It really is wrong that men should be allowed
> to go about loose who fill you with such a strong desire to
> kick them as that little man does.
>
> I send herewith a spare copy of *Nineteenth* with my
> paper about Tyndall. It is not exactly what I could wish,
> as I was hurried over it, and knocked up into the bargain,
> but I have tried to give a fair view of him. Tell me
> what you think of it.
>
> I have been having a day or two on the sick list.
> Nothing discernible the matter, only flopped, as I did in
> the spring. However, I am picking up again. The fact
> is, I have never any blood pressure to spare, and a small
> thing humbugs the pump.

However, I have some kicks left in me, *vide* the preface to the fourth vol. of *Essays;* do. No. V. when that appears in February.

Now, my dear old friend, take care of yourself in the coming year '94. I'll stand by you as long as the fates will let me, and you must be equally "Johnnie." With our love to Lady Hooker and yourself—Ever yours affectionately,   T. H. HUXLEY.

CHAPTER XIII

1894

THE completion early in 1894 of the ninth volume of *Collected Essays* was followed by a review of them in *Nature* (February 1), from the pen of Professor Ray Lankester, emphasising the way in which the writer's personality appears throughout the writing :—

There is probably no lover of apt discourse, of keen criticism, or of scientific doctrine who will not welcome the issue of Professor Huxley's *Essays* in the present convenient shape. For my own part, I know of no writing which by its mere form, even apart from the supreme interest of the matters with which it mostly deals, gives me so much pleasure as that of the author of these essays. In his case, more than that of his contemporaries, it is strictly true that the style is the man. Some authors we may admire for the consummate skill with which they transfer to the reader their thought without allowing him, even for a moment, to be conscious of their personality. In Professor Huxley's work, on the other hand, we never miss his fascinating presence ; now he is gravely shaking his head, now compressing the lips with emphasis, and from time to time, with a quiet twinkle of the eye, making unexpected apologies or protesting that he is of a modest and peace-loving nature.   At the same time, one becomes accustomed to a rare and delightful phenomenon.   Everything which has entered the author's brain by eye or ear, whether of recondite philosophy, biological fact, or political programme, comes out again to us — clarified, sifted, arranged, and vivified by its passage through the logical machine of his strong individuality.

Of the artist in him it continues :—

He deals with form not only as a mechanical engineer *in partibus* (Huxley's own description of himself), but also as an artist, a born lover of form, a character which others recognise in him though he does not himself set it down in his analysis.

The essay on "Animal Automatism" suggested a reminiscence of Professor Lankester's as to the way in which it was delivered, and this in turn led to Huxley's own account of the incident in the letter given in vol. ii. p. 134.

About the same time there is a letter acknowledging Mr. Bateson's book *On Variation*, which is interesting as touching on the latter-day habit of speculation apart from fact which had begun to prevail in biology :—

HODESLEA, *Feb.* 20, 1894.

MY DEAR MR. BATESON—I have put off thanking you for the volume *On Variation* which you have been so good as to send me in the hope that I should be able to look into it before doing so.

But as I find that impossible, beyond a hasty glance, at present, I must content myself with saying how glad I am to see from that glance that we are getting back from the region of speculation into that of fact again.

There have been threatenings of late that the field of battle of Evolution was being transferred to Nephelococcygia.

I see you are inclined to advocate the possibility of considerable "saltus" on the part of Dame Nature in her variations. I always took the same view, much to Mr. Darwin's disgust, and we used often to debate it.

If you should come across my article in the *Westminster* (1860) you will find a paragraph on that question near the end. I am writing to Macmillan to send you the volume.—Yours very faithfully,

T. H. HUXLEY.

By the way, have you ever considered this point, that the variations of which breeders avail themselves are exactly those which occur when the previously wild stocks are subjected to exactly the same conditions?

The rest of the first half of the year is not eventful. As illustrating the sort of communications which constantly came to him, I quote from a letter to Sir J. Donnelly, of January 11 :—

I had a letter from a fellow yesterday morning who must be a lunatic, to the effect that he had been reading my essays, thought I was just the man to spend a month with, and was coming down by the five o'clock train, attended by his seven children and his *mother-in-law!*

Frost being over, there was lots of boiling water ready for him, but he did not turn up !

Wife and servants expected nothing less than assassination.

Later he notes with dismay an invitation as a Privy Councillor to a State evening party :—

It is at 10.30 P.M., just the time this poor old septuagenarian goes to bed !

My swellness is an awful burden, for as it is I am going to dine with the Prime Minister on Saturday.

The banquet with the Prime Minister here alluded to was the occasion of a brief note of apology · to Lord Rosebery for having unintentionally kept him waiting :—

HODESLEA, EASTBOURNE,
May 28, 1894.

DEAR LORD ROSEBERY—I had hoped that my difficulties in dealing with an overtight scabbard stud, as we sat down to dinner on Saturday had inconvenienced no one but myself, until it flashed across my mind after I had parted from you that, as you had observed them, it was only too probable that I had the misfortune to keep you waiting.

I have been in a state of permanent blush ever since, and I feel sure you will forgive me for troubling you with this apology as the only remedy to which I can look for relief from that unwonted affliction.—I am, dear Lord Rosebery, yours very faithfully,   T. H. HUXLEY.

All through the spring he had been busy completing the chapter on Sir Richard Owen's work, which he had been asked to write by the biographer of his old opponent, and on February 4 tells Sir J. D. Hooker :—

I am toiling over my chapter about Owen, and I believe his ghost in Hades is grinning over my difficulties.

The thing that strikes me most is, how he and I and all the things we fought about belong to antiquity.

It is almost impertinent to trouble the modern world with such antiquarian business.

He sent the MS. to Sir M. Foster on June 16; the book itself appeared in December. The chapter in question was restricted to a review of the immense amount of work, most valuable on its positive side, done by Owen (compare the letter of January 18, 1893, p. 273); and the review in *Nature* remarks of it that the criticism is "so straightforward, searching, and honest as to leave nothing further to be desired."

Besides this piece of work, he had written early in the year a few lines on the general character of the nineteenth century, in reply to a request, addressed to "the most illustrious children of the century," for their opinion as to what name will be given to it by an impartial posterity—the century of Comte, of Darwin or Renan, of Edison, Pasteur, or Gladstone. He replied :—

I conceive that the leading characteristic of the nineteenth century has been the rapid growth of the scientific spirit, the consequent application of scientific methods of investigation to all the problems with which the human mind is occupied, and the correlative rejection of traditional beliefs which have proved their incompetence to bear such investigation.

The activity of the scientific spirit has been manifested in every region of speculation and of practice.

Many of the eminent men you mention have been its effective organs in their several departments.

But the selection of any one of these, whatever his merits, as an adequate representative of the power and majesty of the scientific spirit of the age would be a grievous mistake.

Science reckons many prophets, but there is not even a promise of a Messiah.

The unexampled increase in the expenditure of the European states upon their armaments led the Arbitration Alliance this year to issue a memorial urging the Government to co-operate with other Governments in reducing naval and military burdens. Huxley was asked to sign this memorial, and replied to the secretary as follows :—

HODESLEA, EASTBOURNE,
June 21, 1894.

DEAR SIR—I have taken some time to consider the memorial to which you have called my attention, and I regret that I do not find myself able to sign it.

Not that I have the slightest doubt about the magnitude of the evils which accrue from the steady increase of European armaments ; but because I think that this regrettable fact is merely the superficial expression of social forces, the operation of which cannot be sensibly affected by agreements between Governments.

In my opinion it is a delusion to attribute the growth of armaments to the " exactions of militarism." The " exactions of industrialism," generated by international commercial competition, may, I believe, claim a much larger share in prompting that growth. Add to this the French thirst for revenge, the most just determination of the German and Italian peoples to assert their national unity ; the Russian Panslavonic fanaticism and desire for free access to the western seas ; the Papacy steadily fishing in the troubled waters for the means of recovering its lost (I hope for ever lost) temporal possessions and spiritual supremacy ; the " sick man," kept alive only because each of his doctors is afraid of the other becoming his heir.

When I think of the intensity of the perturbing agencies which arise out of these and other conditions of modern European society, I confess that the attempt to counteract them by asking Governments to agree to a maximum military expenditure, does not appear to me to be worth making ; indeed I think it might do harm by leading people to suppose that the desires of Governments are the chief agents in determining whether peace or war shall obtain in Europe.—I am, yours faithfully,

T. H. HUXLEY.

Later in the year, on August 8, took place the meeting of the British Association at Oxford, noteworthy for the presidential address delivered by Lord Salisbury, Chancellor of the University, in which the doctrine of evolution was " enunciated as a matter of course—disputed by no reasonable man," although accompanied by a description of the working of natural selection and variation which appeared to the man of science a mere travesty of these doctrines.

Huxley had been persuaded to attend this meeting, the more willingly, perhaps, since his reception at Oxford the year before suggested that there would be a special piquancy in the contrast between this and the last meeting of the Association at Oxford in 1860. He was not disappointed. Details apart, the cardinal situation was reversed. The genius of the place had indeed altered. The representatives of the party, whose prophet had once contemptuously come here to anathematise the *Origin*, returned at length to the same spot to admit—if not altogether ungrudgingly—the greatness of the work accomplished by Darwin.

Once under promise to go, he could not escape

without the "few words" which he now found so
tiring; but he took the part which assured him
greatest freedom, as seconder of the vote of thanks
to the president for his address. The study of an
advance copy of the address raised an "almost over-
whelming temptation" to criticise certain statements
contained in it; but this would have been out of
place in seconding a vote of thanks; and resisting
the temptation, he only "conveyed criticism," as he
writes to Professor Lewis Campbell, "in the form
of praise": going so far as to suggest "it might be
that, in listening to the deeply interesting address
of the President, a thought had occasionally entered
his mind how rich and profitable might be the
discussion of that paper in Section D" (Biology).
It was not exactly an offhand speech. Writing to
Sir M. Foster for any good report which might
appear in an Oxford paper, he says :—

I have no notes of it. I wrote something on Tuesday
night, but this draft is no good, as it was metamorphosed
two or three times over on Wednesday.

One who was present and aware of the whole
situation once described how he marked the eyes of
another interested member of the audience, who knew
that Huxley was to speak, but not what he meant
to say, turning anxiously whenever the president
reached a critical phrase in the address, to see how
he would take it. But the expression of his face
told nothing; only those who knew him well could

infer a suppressed impatience from a little twitching of his foot.

Of this occasion Professor Henry F. Osborn, one of his old pupils, writes in his "Memorial Tribute to Thomas H. Huxley" (*Transactions of the N.Y. Acad. Soc.* vol. xv.):—

Huxley's last public appearance was at the meeting of the British Association at Oxford. He had been very urgently invited to attend, for, exactly a quarter of a century before, the Association had met at Oxford, and Huxley had had his famous encounter with Bishop Wilberforce. It was felt that the anniversary would be an historic one, and incomplete without his presence, and so it proved to be. Huxley's especial duty was to second the vote of thanks for the Marquis of Salisbury's address—one of the invariable formalities of the opening meetings of the Association. The meeting proved to be the greatest one in the history of the Association. The Sheldonian Theatre was packed with one of the most distinguished scientific audiences ever brought together, and the address of the Marquis was worthy of the occasion. The whole tenor of it was the unknown in science. Passing from the unsolved problems of astronomy, chemistry, and physics, he came to biology. With delicate irony he spoke of the "*comforting word, evolution,*" and passing to the Weismannian controversy, implied that the diametrically opposed views so frequently expressed nowadays threw the whole process of evolution into doubt. It was only too evident that the Marquis himself found no comfort in evolution, and even entertained a suspicion as to its probability. It was well worth the whole journey to Oxford to watch Huxley during this portion of the address. In his red doctor-of-laws gown, placed upon his shoulders by the very body of men who had once referred to him as "a Mr.

Huxley," [1] he sank deeper into his chair upon the very front of the platform and restlessly tapped his foot. His situation was an unenviable one. He had to thank an ex-Prime Minister of England and present Chancellor of Oxford University for an address, the sentiments of which were directly against those he himself had been maintaining for twenty-five years. He said afterwards that when the proofs of the Marquis's address were put into his hands the day before, he realised that he had before him a most delicate and difficult task. Lord Kelvin (Sir William Thomson) one of the most distinguished living physicists, first moved the vote of thanks, but his reception was nothing to the tremendous applause which greeted Huxley in the heart of that University whose cardinal principles he had so long been opposing. Considerable anxiety had been felt by his friends lest his voice should fail to fill the theatre, for it had signally failed during his Romanes Lecture delivered in Oxford the year before, but when Huxley arose he reminded you of a venerable gladiator returning to the arena after years of absence. He raised his figure and his voice to its full height, and, with one foot turned over the edge of the step, veiled an unmistakable and vigorous protest in the most gracious and dignified speech of thanks.

Throughout the subsequent special sessions of this meeting Huxley could not appear. He gave the impression of being aged but not infirm, and no one realised that he had spoken his last word as champion of the law of evolution. [2]

Such criticism of the address as he actually expressed reappears in the leading article, "Past and Present," which he wrote for *Nature* to celebrate the

---

[1] This phrase was actually used by the *Times*.
[2] See, however, p. 342.

twenty-fifth anniversary of its foundation (Nov. 1, 1894).

The essence of the criticism is that with whatever demonstrations of hostility to parts of the Darwinian theory Lord Salisbury covered the retreat of his party from their ancient positions, he admitted the validity of the main points for which Darwin contended.

The essence of this great work (the *Origin of Species*) may be stated summarily thus: it affirms the mutability of species and the descent of living forms, separated by differences of more than varietal value, from one stock. That is to say, it propounds the doctrine of evolution as far as biology is concerned. So far, we have merely a restatement of a doctrine which, in its most general form, is as old as scientific speculation. So far, we have the two theses which were declared to be scientifically absurd and theologically damnable by the Bishop of Oxford in 1860.

It is also of these two fundamental doctrines that, at the meeting of the British Association in 1894, the Chancellor of the University of Oxford spoke as follows :—

" Another lasting and unquestioned effect has resulted from Darwin's work. He has, as a matter of fact, disposed of the doctrine of the immutability of species. . . ."

" Few now are found to doubt that animals separated by differences far exceeding those that distinguished what we know as species have yet descended from common ancestors."

Undoubtedly, every one conversant with the state of biological science is aware that general opinion has long had good reason for making the *volte face* thus indicated. It is also mere justice to Darwin to say that this "lasting and unquestioned" revolution is, in a very real sense, his work. And yet it is also true that, if all the conceptions promulgated in the *Origin of Species* which are peculiarly Darwinian were swept away, the theory of the evolution of animals and plants would not be in the slightest degree shaken.

The strain of this single effort was considerable; "I am frightfully tired," he wrote on August 11, "but the game was worth the candle."

Letters to Sir J. D. Hooker and to Professor Lewis Campbell contain his own account of the affair. The reference in the latter to the priests is in reply to Professor Campbell's story of one of Jowett's last sayings. They had been talking of the collective power of the priesthood to resist the introduction of new ideas; a long pause ensued, and the old man seemed to have slipped off into a doze, when he suddenly broke the silence by saying, "The priests will always be too many for you."

> THE SPA, TUNBRIDGE WELLS,
> *Aug.* 12, 1894.

MY DEAR HOOKER—I wish, as everybody wished, you had been with us on Wednesday evening at Oxford when we settled accounts for 1860, and got a receipt in full from the Chancellor of the University, President of the Association, and representative of ecclesiastical conservatism and orthodoxy.

I was officially asked to second the vote of thanks for the address, and got a copy of it the night before—luckily —for it was a kittle business. . . .

It was very queer to sit there and hear the doctrines you and I were damned for advocating thirty-four years ago at Oxford, enunciated as matters of course—disputed by no reasonable man !—in the Sheldonian Theatre by the Chancellor. . . .

Of course there is not much left of me, and it will take a fortnight's quiet at Eastbourne (whither we return on Tuesday next) to get right.   But it was a pleasant last flare-up in the socket !

With our love to you both—Ever yours affectionately,
T. H. HUXLEY.

HODESLEA, *Aug.* 18, 1894.

MY DEAR CAMPBELL—I am setting you a good example.   You and I are really too old friends to go on wasting ink in honorary prefixes.

I had a very difficult task at Oxford.   The old Adam, of course, prompted the tearing of the address to pieces, which would have been a very easy job, especially the latter half of it.   But as that procedure would not have harmonised well with the function of a seconder of a vote of thanks, and as, moreover, Lord S. was very just and good in his expressions about Darwin, I had to convey criticism in the shape of praise.

It was very curious to me to sit there and hear the Chancellor of the University accept, as a matter of course, the doctrines for which the Bishop of Oxford coarsely anathematised us thirty-four years earlier.   *E pur si muove!*

I am not afraid of the priests in the long-run. Scientific method is the white ant which will slowly but surely destroy their fortifications.   And the importance of scientific method in modern practical life— always growing and increasing—is the guarantee for the gradual emancipation of the ignorant upper and lower classes, the former of whom especially are the strength of the priests.

My wife had a very bad attack of her old enemy some weeks ago, and she thought she would not be able to go to Oxford.   However, she picked up in the wonderfully elastic way she has, and I believe was less

done-up than I when we left on the Friday morning.
I was glad the wife was there, as the meeting gave me
a very kind reception, and it was probably the last flare-
up in the socket.

The Warden of Merton took great care of us, but it
was sad to think of the vacuity of Balliol.

Please remember me very kindly to Father Steffens
and the Steeles, and will you tell Herr Walther we are
only waiting for a balloon to visit the hotel again?

With our affectionate regards to Mrs. Campbell and
yourself—Ever yours very faithfully, T. H. HUXLEY.

Here also belong several letters of miscellaneous
interest. One is to Mrs. Lewis Campbell at the
Maloja :—

HODESLEA, *Aug.* 20, 1894.

MY DEAR MRS. CAMPBELL—What a pity I am not a
telepath! I might have answered your inquiry in the
letter I was writing to your husband yesterday.

The flower I found on the island in Sils Lake was a
cross between *Gentiana lutea* and *Gentiana punctata*—
nothing new, but interesting in many ways as a natural
hybrid.

As to baptizing the island, I am not guilty of usurp-
ing ecclesiastical functions to that extent. I have a
notion that the island has a name already, but I cannot
recollect it. Walther would know.

My wife had a bad attack, and we were obliged to
give up some visits we had projected. But she got well
enough to go to Oxford with me for a couple of days,
and really stood the racket better than I did.

At present she is fairly well, and I hope the enemy
may give her a long respite. The Colliers come to us
at the end of this month, and that will do her good.

With our affectionate regards to you both and remembrances to our friends—Ever yours very truly,

T. H. HUXLEY.

The first of the following set refers to a lively piece of nonsense which Huxley wrote just before going to stay with the Romanes' at Oxford on the occasion of the Romanes Lecture.[1]   After Professor Romanes' death, Mrs. Romanes asked leave to print it in the biography of her husband.   In the other letters, Huxley gives his consent, but, with his usual care for the less experienced, tried to prevent any malicious perversion of the fun which might put her in a false position.

To Mrs. Romanes

HODESLEA, *Sept.* 20, 1894.

I do not think I can possibly have any objection to your using my letter if you think it worth while—but perhaps you had better let me look at it, for I remember nothing about it—and my letters to people whom I trust are sometimes more plain-spoken than polite about things and men.   You know at first there was some talk of my possibly supplying Gladstone's place in case of his failure, and I would not be sure of my politeness in that quarter !

Pray do not suppose that your former letter was other than deeply interesting and touching to me.   I had more than half a mind to reply to it, but hesitated with a man's horror of touching a wound he cannot heal.

And then I got a bad bout of " liver," from which I am just picking up.

HODESLEA, *Sept.* 22, 1894.

It's rather a rollicking epistle, I must say, but as my wife (who sends her love) says she thinks she is the only person who has a right to complain (and she does not), I do not know why it should not be published.

[1] See p. 289.

P.S.—I fancy very few people will catch the allusion about not contradicting me. But perhaps it would be better to take the opinion of some impartial judge on that point.

I do not care the least on my own account, but I see my words might be twisted into meaning that you had told me something about your previous guest, and that I referred to what you had said.

Of course you had done nothing of the kind, but as a wary old fox, experienced sufferer from the dodges of the misrepresenter, I feel bound not to let you get into any trouble if I can help it.

A regular lady's *P.S.* this.

P.S.—Letter returned herewith.

To Mr. Leslie Stephen

Hodeslea, *Oct.* 16, 1894.

My dear Stephen—I am very glad you like to have my *omnium gatherum*, and think the better of it for gaining me such a pleasant letter of acknowledgment.

It is a great loss to me to be cut off from all my old friends, but sticking closely to my hermitage, with fresh air and immense quantities of rest, have become the conditions of existence for me, and one must put up with them.

I have not paid all the debt incurred in my Oxford escapade yet—the last "little bill" being a sharp attack of lumbago, out of which I hope I have now emerged. But my deafness alone should bar me from decent society. I have not the moral courage to avoid making shots at what people say, so as not to bore them ; and the results are sometimes disastrous.

I don't see there is any real difference between us. You are charitable enough to overlook the general immorality of the cosmos on the score of its having begotten morality in one small part of its domain.—Ever yours very faithfully,   T. H. HUXLEY.

To Mr. G—— S——[1]

HODESLEA, *Oct.* 31, 1894.

DEAR MR. S—— " Liver," " lumbago," and other small ills the flesh is heir to, have been making me very lazy lately, especially about letter-writing.

You have got into the depths where the comprehensible ends in the incomprehensible—where the symbols which may be used with confidence so far begin to get shaky.

It does not seem to me absolutely necessary that matter should be composed of solid particles.   The "atoms" may be persistent whirlpools of a continuous "substance"— which substance, if at rest, could not affect us (all sensory impression being dependent on motion) and subsequently would *for us* $= 0$.   The evolution of matter would be the getting under weigh of this "nothing for us" until it became the "something for us," the different motions of which give us the mental states we call the qualities of things.

But it needs a very steady head to walk safely among these abysses of thought, and the only use of letting the mind range among them is as a corrective to the hasty dogmatism of the so-called materialists, who talk just as glibly of that of which they know nothing as the most bigoted of the orthodox.

Here also stand two letters to Lord Farrer, one before, the other after, his address at the Statistical Society on the Relations between Morals, Economics and Statistics, which touch on several philosophical and social questions, always, to his mind, intimately

[1] See p. 310.

connected, and wherein wrong modes of thought indubitably lead to wrong modes of action. Noteworthy is a defence of the fundamental method of Political Economy, however much its limitations might be forgotten by some of its exponents. The reference to the Church agitation to introduce dogmatic teaching into the elementary schools has also a lasting interest.

HODESLEA, Nov. 6, 1894.

My dear Farrer — Whenever you get over the optimism of your youthful constitution (I wish I were endowed with that blessing) you will see that the Gospels and I are right about the Devil being " Prince " (note the distinction—not " king ") of the Cosmos.

The *a priori* road to scientific, political, and all other doctrine is H.R.H. Satan's invention—it is the intellectual, broad, and easy path which leadeth to Jehannum.

The King's road is the strait path of painful observation and experiment, and few they be that enter thereon.

R. G. Latham, queerest of men, had singular flashes of insight now and then. Forty years ago he gravely told me that the existence of the Established Church was to his mind one of the best evidences of the recency of the evolution of the human type from the simian.

How much there is to confirm this view in present public opinion and the intellectual character of those who influence it!

It explains all your difficulties at once, and I regret that I do not seem to have mentioned it at any of those mid-day symposia which were so pleasant when you and I were younger.—Ever yours very faithfully,

T. H. HUXLEY.

P.S.—Apropos of Athelstan Riley and his friends—I feel rather obliged to them. I assented to the compromise

(1) because I felt that English opinion would not let us have the education of the masses at any cheaper price; (2) because, with the Bible in lay hands, I was satisfied that the teaching from it would gradually become modified into harmony with common sense.

I do not doubt that this is exactly what has happened, and is the ground of the alarm of the orthodox.

But I do not repent of the compromise in the least. Twenty years of reasonably good primary education is " worth a mass."

Moreover the Diggleites stand to lose anyhow, and they will lose most completely and finally if they win at the elections this month. So I am rather inclined to hope they may.

HODESLEA, STAVELEY ROAD, EASTBOURNE,
Nov. 3, 1894.

MY DEAR MR. CLODD—They say that the first thing an Englishman does when he is hard up for money is to abstain from buying books. The first thing I do when I am liver-y, lumbagy, and generally short of energy, is to abstain from answering letters. And I am only just emerging from a good many weeks of that sort of flabbiness and poverty.

Many thanks for your notice of Kidd's book. Some vile punsters called it an attempt to put a Kid glove on the iron hand of Nature. I thought it (I mean the book, not the pun) clever from a literary point of view, and worthless from any other. You will see that I have been giving Lord Salisbury a Roland for his Oliver in <i>Nature.</i> But, as hinted, if we only had been in Section D !

With my wife's and my kind regards and remembrances—Ever yours very truly, T. H. HUXLEY.

ATHENÆUM CLUB, Dec. 19, 1894.

MY DEAR FARRER—I am indebted to you for giving the recording angel less trouble than he might otherwise have had, on account of the worse than usual un-punctuality of the London and Brighton this morning. For I have utilised the extra time in reading and thinking over your very interesting address.

Thanks for your protest against the mischievous *a priori* method, which people will not understand is as gross an anachronism in social matters as it would be in Hydrostatics. The so-called "Sociology" is honeycombed with it, and it is hard to say who are worse, the individualists or the collectivists. But in your just wrath don't forget that there is such a thing as a science of social life, for which, if the term had not been so hopelessly degraded, Politics is the proper name.

Men are beings of a certain constitution, who, under certain conditions, will as surely tend to act in certain ways as stones will tend to fall if you leave them unsupported. The laws of their nature are as invariable as the laws of gravitation, only the applications to particular cases offer worse problems than the case of the three bodies.

The Political Economists have gone the right way to work—the way that the physical philosopher follows in all complex affairs—by tracing out the effects of one great cause of human action, the desire of wealth, supposing it to be unchecked.

If they, or other people, have forgotten that there are other potent causes of action which may interfere with this, it is no fault of scientific method but only their own stupidity.

Hydrostatics is not a "dismal science," because water does not always seek the lowest level *e.g.* from a bottle turned upside down, if there is a cork in the neck !

There is much need that somebody should do for what is vaguely called "Ethics" just what the Political Economists have done. Settle the question of what will be done under the unchecked action of certain motives, and leave the problem of "ought" for subsequent consideration.

For, whatever they ought to do, it is quite certain the majority of men will act as if the attainment of certain positive and negative pleasures were the end of action.

We want a science of "Eubiotics" to tell us exactly what will happen if human beings are exclusively actuated by the desire of well-being in the ordinary sense. Of course the utilitarians have laid the foundations of such a science, with the result that the nicknamer of genius called this branch of science "pig philosophy," making just the same blunder as when he called political economy "dismal science."

"Moderate well-being" may be no more the worthiest end of life than wealth. But if it is the best to be had in this queer world—it may be worth trying for.

But you will begin to wish the train had been *punctual!*

Draw comfort from the fact that if error is always with us, it is, at any rate, remediable. I am more hopeful than when I was young. Perhaps life (like matrimony, as some say) should begin with a little aversion!—Ever yours very faithfully,   T. H. HUXLEY.

Some years before this, a fund for a "Darwin Medal" had been established in memory of the great naturalist, the medal to be awarded biennially for researches in biology. With singular appropriateness, the first award was made to Dr. A. R. Wallace, the joint propounder of the theory of Natural Selection, whose paper, entrusted to Darwin's literary sponsorship, caused the speedy publication of Darwin's own long-continued researches and speculations. The second, with equal appropriateness, was to Sir J. D. Hooker, both as a leader in science and a helper and adviser of Darwin.

Huxley's own view of such scientific honours as medals and diplomas was that they should be employed to stimulate for the future rather than to reward for the past; and delighted as he was at the poetic justice of these two awards, this justice once satisfied, he let his opinion be known that thenceforward the Darwin Medal ought to be given only to younger men. But when this year he found the Darwin Medal awarded to himself "for his researches in biology and his long association with Charles Darwin," he could not but be touched and gratified by this mark of appreciation from his fellow-workers in science, this association in one more scientific record with old allies and true friends—to "have his niche in the Pantheon" next to Hooker and near to Darwin.

It was a rare instance of the fitness of things that the three men who had done most to develop and to defend Darwin's ideas should live to stand first in the list of the Darwin medalists; and Huxley felt this to be a natural closing of a chapter in his life, a fitting occasion on which to bid farewell to public life in the world of science. Almost at the same moment another chapter in science reached its completion in the "coming of age" of *Nature*, a journal which, when scientific interests at large had grown stronger, had succeeded in realising his own earlier efforts to found a scientific organ, and with which he had always been closely associated.

As mentioned above, he wrote for the November

number an introductory article called "Past and Present," comparing the state of scientific thought of the day with that of twenty-five years before, when the journal was first started. To celebrate the occasion, a dinner was to be held this same month of all who had been associated with *Nature*, and this Huxley meant to attend, as well as the more important anniversary dinner of the Royal Society on St. Andrew's Day.

I have promised (he writes on November 6 to Sir M. Foster) to go to the *Nature* dinner if I possibly can. Indeed I should be sorry to be away. As to the R.S. nothing short of being confined to bed will stop me. And I shall be good for a few words after dinner.

Thereafter I hope not to appear again on any stage.

His letter about the medal expresses his feelings as to the award.

HODESLEA, *Nov.* 2, 1894.

MY DEAR FOSTER—Didn't I tell the P.R.S., Secretaries, Treasurer, and all the Fellows thereof, when I spake about Hooker years ago, that thenceforth the Darwin Medal was to be given to the young, and not to useless old extinct volcanoes? I ought to be very angry with you all for coolly ignoring my wise counsels.

But whether it is vanity or something a good deal better, I am not. One gets chill old age, and it is very pleasant to be warmed up unexpectedly even against one's injunctions. Moreover, my wife is very pleased, not to say jubilant; and if I were made Archbishop of Canterbury I should not be able to convince her that my services to Theology were hardly of the sort to be rewarded in that fashion.

I need not say what I think about your action in the matter, my faithful old friend.   With our love to you both—Ever yours,   T. H. HUXLEY.

I suppose you are all right again, as you write from the R.S.   Liver permitting I shall attend meeting and dinner.   It is very odd that the Medal should come along with my pronouncement in *Nature*, which I hope you like.   I cut out rather a stinging paragraph at the end.

HODESLEA, EASTBOURNE,
Nov. 11, 1894.

MY DEAR DONNELLY—Why on earth did I not answer your letter before ?   Echo (being Irish) says, " Because of your infernal bad habit of putting off ; which is growing upon you, you wretched old man."

Of course I shall be very glad if anything can be done for S——.   Howes has written to me about him since your letter arrived—and I am positively going to answer his epistle.   It's Sunday morning, and I feel good.

You will have seen that the R.S. has been giving me the Darwin Medal, though I gave as broad a hint as was proper the last time I spoke at the Anniversary, that it ought to go to the young men.   Nevertheless, with ordinary inconsistency of the so-called " rational animal," I am well pleased.

I hope you will be at the dinner, and would ask you to be my guest—but as I thought my boys and boys-in-law would like to be there, I have already exceeded my lawful powers of invitation, and had to get a dispensation from Michael Foster.

I suppose I shall be like a horse that " stands at livery " for some time after—but it is positively my last appearance on any stage.

We were very glad to hear from Lady Donnelly that you had had a good and effectual holiday.   With our love—Ever yours,   T. H. HUXLEY.

I return Howes' letter in case you want it. I see I
need not write to him again after all.   Three cheers !

Please give Lady Donnelly this.   A number of estimable members of her sex have flown at me for writing
what I thought was a highly complimentary letter.   But
she will be just, I know.

" The best of women are apt to be a little weak in the
great practical arts of give-and-take, and putting up with
a beating, and a little too strong in their belief in the
efficacy of government.   Men learn about these things in
the ordinary course of their business ; women have no
chance in home life, and the boards and councils will be
capital schools for them.   Again, in the public interest
it will be well ; women are more naturally economical
than men, and have none of our false shame about looking
after pence.   Moreover, they don't job for any but their
lovers, husbands, and children, so that we know the
worst."

The speech at the Royal Society Anniversary
dinner—which he evidently enjoyed making—was a
fine piece of speaking, and quite carried away the
audience, whether in the gentle depreciation of his
services to science, or in his profession of faith in the
methods of science and the final triumph of the
doctrine of evolution, whatever theories of its operation might be adopted or discarded in the course of
further investigation.

I quote from the *Times* report of the speech :—

But the most difficult task that remains is that which
concerns myself.   It is 43 years ago this day since the
Royal Society did me the honour to award me a Royal
medal, and thereby determined my career.   But, having
long retired into the position of a veteran, I confess that

I was extremely astonished—I honestly also say that I was extremely pleased to receive the announcement that you had been good enough to award to me the Darwin Medal.   But you know the Royal Society, like all things in this world, is subject to criticism.   I confess that with the ingrained instincts of an old official that which arose in my mind after the reception of the information that I had been thus distinguished was to start an inquiry which I suppose suggests itself to every old official—How can my Government be justified ?   In reflecting upon what had been my own share in what are now very largely ancient transactions, it was perfectly obvious to me that I had no such claims as those of Mr. Wallace.   It was perfectly clear to me that I had no such claims as those of my lifelong friend Sir Joseph Hooker, who for 25 years placed all his great sources of knowledge, his sagacity, his industry, at the disposition of his friend Darwin. And really, I begin to despair of what possible answer could be given to the critics whom the Royal Society, meeting as it does on November 30, has lately been very apt to hear about on December 1.   Naturally there occurred to my mind that famous and comfortable line, which I suppose has helped so many people under like circumstances, " They also serve who only stand and wait." I am bound to confess that the standing and waiting, so far as I am concerned, to which I refer, has been of a somewhat peculiar character.   I can only explain it, if you will permit me to narrate a story which came to me in my old nautical days, and which, I believe, has just as much foundation as a good deal of other information which I derived at the same period from the same source. There was a merchant ship in which a member of the Society of Friends had taken passage, and that ship was attacked by a pirate, and the captain thereupon put into the hands of the member of the Society of Friends a pike, and desired him to take part in the subsequent action, to which, as you may imagine, the reply was that he would do nothing of the kind; but he said that he had no objection to stand and wait at the gangway. He did stand and wait with the pike in his hands, and when the pirates mounted and showed themselves coming on board he thrust his pike with the sharp end forward into the persons who were mounting, and he said, "Friend, keep on board thine own ship." It is in that sense that I venture to interpret the principle of standing and waiting to which I have referred. I was convinced as firmly as I have ever been convinced of anything in my life, that the *Origin of Species* was a ship laden with a cargo of rich value, and which, if she were permitted to pursue her course, would reach a veritable scientific Golconda, and I thought it my duty, however naturally averse I might be to fighting, to bid those who would disturb her beneficent operations to keep on board their own ship. If it has pleased the Royal Society to recognise such poor services as I may have rendered in that capacity, I am very glad, because I am as much convinced now as I was 34 years ago that the theory propounded by Mr. Darwin—I mean that which he propounded, not that which has been reported to be his by too many ill-instructed, both friends and foes—has never yet been shown to be inconsistent with any positive observations, and if I may use a phrase which I know has been objected to, and which I use in a totally different sense from that in which it was first proposed by its first propounder, I do believe that on all grounds of pure science it "holds the field," as the only hypothesis at present before us which has a sound scientific foundation. It is quite possible that you will apply to me the remark that has often been applied to persons in such a position as mine, that we are apt to exaggerate the importance of that to which our lives have been more or less devoted. But I am sincerely of opinion that the views which were propounded by Mr. Darwin 34 years ago may be understood hereafter as constituting an epoch in the intellectual history of the human race. They will modify the whole system of our thought and opinion, our most intimate convictions. But I do not know, I do not think anybody knows, whether the particular views which he held will be hereafter fortified by the experience of the ages which come after us; but of this thing I am perfectly certain, that the present course of things has resulted from the feeling of the smaller men who have followed him that they are incompetent to bend the bow of Ulysses, and in consequence many of them are seeking their salvation in mere speculation. Those who wish to attain to some clear and definite solution of the great problems which Mr. Darwin was the first person to set before us in later times must base themselves upon the facts which are stated in his great work, and, still more, must pursue their inquiries by the methods of which he was so brilliant an exemplar throughout the whole of his life. You must have his sagacity, his untiring search after the knowledge of fact, his readiness always to give up a preconceived opinion to that which was demonstrably true, before you can hope to carry his doctrines to their ultimate issue; and whether the particular form in which he has put them before us may be such as is finally destined to survive or not is more, I venture to think, than anybody is capable at this present moment of saying. But this one thing is perfectly certain—that it is only by pursuing his methods, by that wonderful single-mindedness, devotion to truth, readiness to sacrifice all things for the advance of definite knowledge, that we can hope to come any nearer than we are at present to the truths which he struggled to attain.

To Sir J. D. Hooker

HODESLEA, EASTBOURNE,
Dec. 4, 1894.

MY DEAR OLD MAN—See the respect I have for your six years' seniority! I wished you had been at the dinner, but was glad you were not. Especially as next morning there was a beastly fog, out of which I bolted home as fast as possible.

I shall have to give up these escapades. They knock me up for a week afterwards. And really it is a pity, just as I have got over my horror of public speaking, and find it very amusing. But I suppose I should gravitate into a bore as old fellows do, and so it is as well I am kept out of temptation.

I will try to remember what I said at the *Nature* dinner.[1] I scolded the young fellows pretty sharply for their slovenly writing.

There will be a tenth vol. of Essays some day, and an Index rerum. Do you remember how you scolded me for being too speculative in my maiden lecture on Animal Individuality forty odd years ago ? " On revient toujours," or, to put it another way, " The dog returns to his etc. etc."

So I am deep in philosophy, grovelling through Diogenes Laertius—Plutarch's *Placita* and sich—and often wondering whether the schoolmasters have any better ground for maintaining that Greek is a finer language than English than the fact that they can't write the latter dialect.

So far as I can see, my faculties are as good (including memory for anything that is not useful) as they were fifty years ago, but I can't work long hours, or live out of fresh air. Three days of London bowls me over.

I expect you are in much the same case. But you seem to be able to stoop over specimens in a way impossible to me. It is that incapacity has made me give up dissection and microscopic work. I do a lot on my back, and I can tell you that the latter posture is an immense economy of strength. Indeed, when my heart

---

[1] A brief report of this speech is to be found in the *British Medical Journal* for December 8, 1894, p. 1262.

was troublesome, I used to spend my time either in active outdoor exercise or horizontally.

The Stracheys were here the other day, and it was a great pleasure to us to see them. I think he has had a very close shave with that accident. There is nobody whom I should more delight to honour—a right good man all round—but I am not competent to judge of his work. You are, and I do not see why you should not suggest it. I would give him a medal for being R. Strachey, but probably the Council would make difficulties.

By the way, do you see the *Times* has practically climbed down about the R.S.—came down backwards like a bear, growling all the time? I don't think we shall have any more first of December criticisms.

Lord help you through all this screed. With our love to you both—Ever yours affectionately,

T. H. HUXLEY.

> Abram, Abraham became
> By will divine ;
> Let pickled Brian's name
> Be changed to Brine ![1]
> *Poetae Minores.*

Poor Brian.—Brutal jest !

The following was written to a friend who had alluded to his painful recollection of a former occasion when he was Huxley's guest at the anniversary dinner of the Royal Society, and was hastily summoned from it to find his wife dying.

I fully understand your feeling about the R.S. Dinner. I have not forgotten the occasion when you were my guest : still less my brief sight of you when I called the next day.

---

[1] Sir Joseph's son, Brian, had fallen into a pan of brine.

These things are the " lachrymae rerum "—the abysmal griefs hidden under the current of daily life, and seemingly forgotten, till now and then they come up to the surface —a flash of agony—like the fish that jumps in a calm pool.

One has one's groan and goes to work again.

If I knew of anything else for it, I would tell you ; but all my experience ends in the questionable thanksgiving, " It's lucky it's no worse."

With which bit of practical philosophy, and our love, believe me, ever yours affectionately,

T. H. HUXLEY.

Before speaking of his last piece of work, in the vain endeavour to complete which he exposed himself to his old enemy, influenza, I shall give several letters of miscellaneous interest.

The first is in reply to Lord Farrer's inquiry as to where he could obtain a fuller account of the subject tersely discussed in the chapter he had contributed to the *Life of Owen*.[1]

HODESLEA, Jan. 26, 1895.

MY DEAR FARRER—Miserable me ! Having addressed myself to clear off a heap of letters that have been accumulating, I find I have not answered an inquiry of yours of nearly a month's standing. I am sorry to say that I cannot tell you of any book (readable or otherwise) that will convert my " pemmican " into decent broth for you.

---

[1] " Which," wrote Lord Farrer, "is just what I wanted as an outline of the Biological and Morphological discussion of the last 100 years. But it is 'Pemmican' to an aged and enfeebled digestion. Is there such a thing as a diluted solution of it in the shape of any readable book ? "

There are histories of zoology and of philosophical anatomy, but they all of them seem to me to miss the point (which you have picked out of the pemmican). Indeed, that is just why I took such a lot of pains over these 50 or 60 pages. And I am immensely tickled by the fact that among all the critical notices I have seen, not a soul sees what I have been driving at as you have done. I really wish you would write a notice of it, just to show these Gigadibses (*vide* Right Rev. Blougram)[1] what blind buzzards they are !

Enter a maid. " Please sir, Mrs. Huxley says she would be glad if you would go out in the sun." "All right, Allen." Anecdote for your next essay on Government !

The fact is, I have been knocked up ever since Tuesday, when our University Deputation came off ; and my good wife (who is laid up herself) suspects me (not without reason) of failing to take advantage of a gleam of sunshine.

By the way, can you help us over the University business ? Lord Rosebery is favourable, and there is absolutely nobody on the other side except sundry Philistines, who, having got their degrees, are desirous of inflating their market value.—Yours very truly,

T. H. HUXLEY.

The next is in answer to an appeal for a subscription, from the Church Army.

Jan. 26, 1895.

I regret that I am unable to contribute to the funds of the Church Army.

I hold it to be my duty to do what I can for the cases of distress of which I have direct knowledge ; and I am

---

[1] See Browning's " Bishop Blougram's Apology " :—"Gigadibs the literary man " with his
Abstract intellectual plan of life
Quite irrespective of life's plainest laws.

glad to be able now and then to give timely aid to the industrious and worthy people with whom, as a householder, I am brought into personal relation; and who are so often engaged in a noiseless and unpitied but earnest struggle to do well.

In my judgment, a domestic servant, who is perhaps giving half her wages to support her old parents, is more worthy of help than half-a-dozen Magdalens.

Under these circumstances, you will understand that such funds as are at my disposal are already fully engaged.

The following is to a gentleman—an American, I think—who sent him a long manuscript, an extraordinary farrago of nonsense, to read and criticise, and help to publish. But as he seemed to have acted in sheer simplicity, he got an answer:—

HODESLEA, *Jan.* 31, 1895.

DEAR SIR—I should have been glad if you had taken the ordinary, and, I think, convenient course of writing for my permission before you sent the essay which has reached me, and which I return by this post. I should then have had the opportunity of telling you that I do not undertake to read, or take any charge of such matters, and we should both have been spared some trouble.

I the more regret this, since being unwilling to return your work without examination, I have looked at it, and feel bound to give you the following piece of advice, which I fear may be distasteful, as good counsel generally is.

Lock up your essay. For two years—if possible, three —read no popular expositions of science, but devote yourself to a course of sound *practical* instruction in elementary physics, chemistry, and biology.

Then re-read your essay; do with it as you think

best ; and, if possible, regard a little more kindly than you are likely to do at present, yours faithfully,

T. H. Huxley.

The following passage from a letter to Sir J. D. Hooker refers to a striking discovery made by Dubois :—

Hodeslea, Eastbourne,
Feb. 14, 1895.

The Dutchmen seem to have turned up something like the "missing link" in Java, according to a paper I have just received from Marsh. I expect he was a Socratic party, with his hair rather low down on his forehead and warty cheeks.

Pithecanthropus erectus Dubois (fossil)

rather Aino-ish about the body, small in the calf, and cheese-cutting in the shins. Le voici !

CHAPTER XIV

1895

Two months of almost continuous frost, during which the thermometer fell below zero, marked the winter of 1894-95. Tough, if not strong, as Huxley's constitution was, this exceptional cold, so lowering to the vitality of age, accentuated the severity of the illness which followed in the train of influenza, and at last undermined even his powers of resistance.

But until the influenza seized him, he was more than usually vigorous and brilliant. He was fatigued, but not more so than he expected, by attending a deputation to the Prime Minister in the depth of January, and delivering a speech on the London University question ; and in February he was induced to write a reply to the attack upon agnosticism contained in Mr. Arthur Balfour's *Foundations of Belief*. Into this he threw himself with great energy, all the more because the notices in the daily press were likely to give the reading public a wrong impression as to its polemic against his own position. Mr. Wilfrid

Ward gives an account of a conversation with him on this subject :—

Some one had sent me Mr. A. J. Balfour's book on the *Foundations of Belief* early in February 1895. We were very full of it, and it was the theme of discussion on the 17th of February, when two friends were lunching with us. Not long after luncheon, Huxley came in, and seemed in extraordinary spirits. He began talking of Erasmus and Luther, expressing a great preference for Erasmus, who would, he said, have impregnated the Church with culture, and brought it abreast of the thought of the times, while Luther concentrated attention on individual mystical doctrines. " It was very trying for Erasmus to be identified with Luther, from whom he differed absolutely. A man ought to be ready to endure persecution for what he does hold ; but it is hard to be persecuted for what you don't hold." I said that I thought his estimate of Erasmus's attitude towards the Papacy coincided with Professor R. C. Jebb's. He asked if I could lend him Jebb's Rede Lecture on the subject. I said that I had not got it at hand, but I added, " I can lend you another book, which I think you ought to read—Balfour's *Foundations of Belief*."

He at once became extremely animated, and spoke of it as those who have read his criticisms, published in the following month, would expect. "You need not lend me that. I have exercised my mind with it a good deal already. Mr. Balfour ought to have acquainted himself with the opinions of those he attacks. One has no objection to being abused for what one *does* hold, as I said of Erasmus ; at least, one is prepared to put up with it. An attack on us by some one who understood our position would do all of us good—myself included. But Mr. Balfour has acted like the French in 1870 : he has gone to war without any ordnance maps, and without having surveyed the scene of the campaign. No human being holds the opinions he speaks of as 'Naturalism.' He is a good debater. He knows the value of a word. The word 'Naturalism' has a bad sound and unpleasant associations. It would tell against us in the House of Commons, and so it will with his readers. 'Naturalism' contrasts with 'supernaturalism.' He has not only attacked us for what we don't hold, but he has been good enough to draw out a catechism for 'us wicked people,' to teach us what we *must* hold."

It was rather difficult to get him to particulars, but we did so by degrees. He said, "Balfour uses the word *phenomena* as applying simply to the outer world and not to the inner world. The only people his attack would hold good of would be the Comtists, who deny that psychology is a science. They may be left out of account. They advocate the crudest eighteenth-century materialism. All the empiricists, from Locke onwards, make the observation of the phenomena of the mind itself quite separate from the study of mere sensation. No man in his senses supposes that the sense of beauty, or the religious feelings (this with a courteous bow to a priest who was present), or the sense of moral obligation, are to be accounted for in terms of sensation, or come to us through sensation." I said that, as I understood it, I did not think Mr. Balfour supposed they would acknowledge the position he ascribed to them, and that one of his complaints was that they did not work out their premises to their logical conclusions. I added that so far as one of Mr. Balfour's chief points was concerned—the existence of the external world —Mill was almost the only man on their side in this century who had faced the problem frankly, and he had been driven to say that all men can know is that there are "permanent possibilities of sensation." He did not seem inclined to pursue the question of an external world, but said that though Mill's "Logic" was very good, empiricists were not bound by all his theories.

He characterised the book as a very good and even

brilliant piece of work from a literary point of view; but as a helpful contribution to the great controversy, the most disappointing he had ever read. I said, "There has been no adverse criticism of it yet." He answered with emphasis, "No! *but there soon will be.*" "From you?" I asked. "I let out no secrets," was the reply.

He then talked with great admiration and affection of Mr. Balfour's brother, Francis. His early death, and W. K. Clifford's (Huxley said), had been the greatest loss to science—not only in England, but in the world—in our time. "Half a dozen of us old fogies could have been better spared." He remembered Frank Balfour as a boy at [Harrow] and saw his unusual talent there. "Then my friend, Michael Foster, took him up at Cambridge, and found out that he had real genius for biology. I used to say there was science in the blood, but this new book of his brother's," he added, smiling, "shows I was wrong."

Apropos to his remark about the Comtists, one of the company pointed out that in later life Comte recognised a science of "the individual," equivalent to what Huxley meant by psychology. "That," he replied, "was due to the influence of Clotilde de Vaux. You see," he added, with a kind of Sir Charles Grandison bow to my wife, "what power your sex may have." As Huxley was going out of the house, I said to him that Father A. B. (the priest who had been present) had not expected to find himself in his company. "No! I trust he had plenty of holy water with him," was the reply.

. . . After he had gone, we were all agreed as to the extraordinary vigour and brilliancy he had shown. Some one said, "He is like a man who is what the Scotch call 'fey.'" We laughed at the idea, but we naturally recalled the remark later on.

The story of how the article was written is told in the following letters. It was suggested by Mr.

Knowles, and undertaken after perusal of the review of the book in the *Times*. Huxley intended to have the article ready for the March number of the *Nineteenth Century*, but it grew longer than he had meant it to be, and partly for this reason, partly for fear lest the influenza, then raging at Eastbourne, might prevent him from revising the whole thing at once, he divided it into two instalments. He writes to one daughter on March 1 :—

I suppose my time will come; so I am "making hay while the sun shines" (in point of fact it is raining and blowing a gale outside) and finishing my counterblast to Balfour before it does come.

Love to all you poor past snivellers from an expectant sniveller.

And to another : —

I think the cavalry charge in this month's *Nineteenth* will amuse you. The heavy artillery and the bayonets will be brought into play next month.

Dean Stanley told me he thought being made a bishop destroyed a man's moral courage. I am inclined to think that the practice of the methods of political leaders destroys their intellect for all serious purposes.

No sooner was the first part safely sent off than the contingency he had feared came to pass; only, instead of the influenza meaning incapacity for a fortnight, an unlucky chill brought on bronchitis and severe lung trouble.[1] The second part of the article was never fully revised for press.

---

[1] As he wrote on February 28 to Sir M. Foster: "If I could compound for a few hours' neuralgia, I would not mind; but

HODESLEA, EASTBOURNE,
February 8, 1895.

MY DEAR KNOWLES—Your telegram came before I had looked at to-day's *Times* and the article on Balfour's book, so I answered with hesitation.

Now I am inclined to think that the job may be well worth doing, in that it will give me the opportunity of emphasising the distinction between the view I hold and Spencer's, and perhaps of proving that Balfour is an agnostic after my own heart. So please send the book.

Only if this infernal weather, which shrivels me up soul and body, lasts, I do not know how long I may be over the business. However, you tell me to take my own time.—Ever yours very faithfully,

T. H. HUXLEY.

HODESLEA, EASTBOURNE,
February 18, 1895.

MY DEAR KNOWLES—I send you by this post an instalment (the larger moiety) of my article, which I should be glad to have set up at once *in slip*, and sent to me as speedily as may be. The rest shall follow in the course of the next two or three days.

I am rather pleased with the thing myself, so it is probably not so very good ! But you will judge for yourself.—Ever yours very faithfully, T. H. HUXLEY.

HODESLEA, EASTBOURNE,
February 19, 1895.

MY DEAR KNOWLES—We send our best congratulations to Mrs. Knowles and yourself on the birth of a grand-

---

those long weeks of debility make me very shy of the influenza demon. Here we are practically isolated. . . . I once asked Gordon why he didn't have the African fever. 'Well,' he said, 'you see, fellows think they shall have it, and they do. I didn't think so, and didn't get it.' Exercise your thinking faculty to that extent."

daughter. I forget whether you have had any previous experience of the " Art d'etre Grandpere " or not—but I can assure you, from 14 such experiences, that it is easy and pleasant of acquirement, and that the objects of it are veritable "articles de luxe," involving much amusement and no sort of responsibility on the part of the possessor.

You shall have the rest of my screed by to-morrow's post.—Ever yours very faithfully,   T. H. HUXLEY.

HODESLEA, EASTBOURNE,
February 20, 1895.

MY DEAR KNOWLES—Seven mortal hours have I been hard at work this day to try to keep my promise to you, and as I find that impossible, I have struck work and will see Balfour and his *Foundations,* and even that ark of literature the *Nineteenth,* at Ballywack, before I do any more.

But the whole affair shall be sent by a morning s post to-morrow. I have the proofs. I have found the thing getting too long for one paper, and requiring far more care than I could put into the next two days—so I propose to divide it, if you see no objection.

And there is another reason for this course. Influenza is raging here. I hear of hundreds of cases, and if it comes my way, as it did before, I go to bed and stop there—"the world forgetting and by the world forgot" —until I am killed or cured. So you would not get your article.

As it stands, it is not a bad gambit. We will play the rest of the game afterwards, D.V. and K.V

Hope mother and baby are doing well.—Ever yours,

T. H. HUXLEY.

HODESLEA, EASTBOURNE,
February 23, 1895, 12.30 P.M.

MY DEAR KNOWLES—I have just played and won as hard a match against time as I ever knew in the days of my youth. The proofs, happily, arrived by the first post, so I got to work at them before 9, polished them off by 12, and put them into the post (myself) by 12.5. So you ought to have them by 6 P.M. And, to make your mind easy, I have just telegraphed to you to say so. But, Lord's sake! let some careful eye run over the part of which I have had no revise—for I am "capable de tout" in the way of overlooking errors.

I am very glad you like the thing. The second instalment shall be no worse.

I grieve to say that my estimation of Balfour, as a thinker, sinks lower and lower, the further I go.

God help the people who think his book an important contribution to thought! The Gigadibsians who say so are past divine assistance!

We are very glad to hear the grandchild and mother are getting on so well.—Ever yours very truly,

T. H. Huxley.

Hodeslea, Eastbourne,
March 8, 1895.

My dear Knowles—The proofs have just arrived, but I am sorry to say that (I believe for the first time in our transactions) I shall have to disappoint you.

Just after I had sent off the MS. influenza came down upon me with a swoop. I went to bed and am there still, with no chance of quitting it in a hurry. My wife is in the same case; *item* one of the maids. The house is a hospital, and by great good fortune we have a capital nurse.

Doctor says it's a mild type,[1] in which case I wonder what severe types may be like. I find coughing continuously for fourteen hours or so a queer kind of mildness.

---

[1] "But in the matter of aches and pains, restless paroxysms of coughing and general incapacity, I can give it a high character for efficiency." (To M. Foster, March 7.)

Could you put in an excuse on account of in-fluenza?

Can't write any more.—Ever yours,   T. H. H.

HODESLEA, EASTBOURNE,
March 19, 1895.

MY DEAR KNOWLES—I am making use of the pen of my dear daughter and good nurse, in the first place to thank you for your cheque, in the second place to say that you must not look for the article this month. I haven't been out of bed since the 1st, but they are fighting a battle with bronchitis over my body.—Ever yours very faithfully,   For T. H. H.,
SOPHY HUXLEY.

The next four months were a period of painful struggle against disease, borne with a patience and gentleness which was rare even in the long experience of the trained nurses who tended him. To natural toughness of constitution he added a power of will unbroken by the long strain; and for the sake of others to whom his life meant so much, he wished to recover and willed to do everything towards recovery. And so he managed to throw off the influenza and the severe bronchitis which attended it. What was marvellous at his age, and indeed would scarcely have been expected in a young man, most serious mischief induced by the bronchitis disappeared. By May he was strong enough to walk from the terrace to the lawn and his beloved saxifrages, and to re-mount the steps to the house without help.

But though the original attack was successfully thrown off, the lung trouble had affected the heart;

and in his weakened state, renal mischief ensued. Yet he held out splendidly, never giving in, save for one hour of utter prostration, all through this weary length of sickness. His first recovery strengthened him in expecting to get well from the second attack. And on June 10 he writes brightly enough to Sir J. D. Hooker :—

HODESLEA, EASTBOURNE,
June 10, 1895.

MY DEAR OLD FRIEND—It was cheering to get your letter and to hear that you had got through winter and diphtheria without scathe.

I can't say very much for myself yet, but I am carried down to a tent in the garden every day, and live in the fresh air all I can. The thing that keeps me back is an irritability of the stomach tending to the rejection of all solid food. However, I think I am slowly getting the better of it—thanks to my constitutional toughness and careful nursing and dieting.

What has Spencer been trampling on the " Pour le merite " for, when he accepted the Lyncei ? I was just writing to congratulate him when, by good luck, I saw he had refused !

The beastly nausea which comes on when I try to do anything warns me to stop.

With our love to you both—Ever yours,
T. H. HUXLEY.

The last time I saw him was on a visit to Eastbourne from June 22-24. I was astonished to find how well he looked in spite of all ; thin, indeed, but browned with the endless sunshine of the 1895 summer as he sat every day in the verandah. His voice was still fairly strong ; he was delighted to see us about him, and was cheerful, even merry at times. As the nurse said, she could not expect him to recover, but he did not look like a dying man. When I asked him how he was, he said, "A mere carcass, which has to be tended by other people." But to the last he looked forward to recovery. One day he told the nurse that the doctors must be wrong about the renal mischief, for if they were right, he ought already to be in a state of coma. This was precisely what they found most astonishing in his case; it seemed as if the mind, the strong nervous organisation, were triumphing over the shattered body. Herein lay one of the chief hopes of ultimate recovery.

As late as June 26 he wrote, with shaky handwriting but indomitable spirit, to relieve his old friend from the anxiety he must feel from the newspaper bulletins.

HODESLEA, EASTBOURNE,
June 26, 1895.

MY DEAR HOOKER—The pessimistic reports of my condition which have got into the papers may be giving you unnecessary alarm for the condition of your old comrade. So I send a line to tell you the exact state of affairs.

There is kidney mischief going on—and it is accompanied by very distressing attacks of nausea and vomiting, which sometimes last for hours and make life a burden.

However, strength keeps up very well considering, and of course all depends upon how the renal business goes. At present I don't feel at all like "sending in my checks," and without being over sanguine I rather incline to think that my native toughness will get the best of it—albuminuria or otherwise.—Ever your faithful friend,

T. H. H.

Misfortunes never come single. My son-in-law, Eckersley, died of yellow fever the other day at San Salvador—just as he was going to take up an appointment at Lima worth £1200 a year. Rachel and her three children have but the slenderest provision.

The next two days there was a slight improvement, but on the third morning the heart began to fail. The great pain subdued by anæsthetics, he lingered on about seven hours, and at half-past three on June 29 passed away very quietly.

He was buried at Finchley, on July 4, beside his brother George and his little son Noel, under the shadow of the oak, which had grown up into a stately young tree from the little sapling it had been when the grave of his first-born was dug beneath it, five and thirty years before.

The funeral was of a private character. An old friend, the Rev. Llewelyn Davies, came from Kirkby Lonsdale to read the service ; the many friends who gathered at the grave-side were there as friends mourning the death of a friend, and all touched with the same sense of personal loss.

By his special direction, three lines from a poem written by his wife, were inscribed upon his tombstone—lines inspired by his own robust conviction that, all question of the future apart, this life as it can be lived, pain, sorrow, and evil notwithstanding, is worth—and well worth—living :—

> Be not afraid, ye waiting hearts that weep ;
> For still He giveth His beloved sleep,
> And if an endless sleep He wills, so best.

CHAPTER XV

HE had intellect to comprehend his highest duty distinctly, and force of character to do it ; which of us dare ask for a higher summary of his life than that ?

Such was Huxley's epitaph upon Henslow ; it was the standard which he endeavoured to reach in his own life. It is the expression of that passion for veracity which was perhaps his strongest characteristic ; an uncompromising passion for truth in thought, which would admit no particle of self-deception, no assertion beyond what could be verified ; for truth in act, perfect straightforwardness and sincerity, with complete disregard of personal consequences for uttering unpalatable fact.

Truthfulness, in his eyes, was the cardinal virtue, without which no stable society can exist. Conviction, sincerity, he always respected, whether on his own side or against him. Clever men, he would say, are as common as blackberries ; the rare thing is to find a good one. The lie from interested motives was only more hateful to him than the lie from self-delusion or foggy thinking. With this he classed the "sin of faith," as he called it; that form of credence which does not fulfil the duty of making a right use of reason; which prostitutes reason by giving assent to propositions which are neither self-evident nor adequately proved.

This principle has always been far from finding universal acceptance. One of his theological opponents went so far as to affirm that a doctrine may be not only held, but dogmatically insisted on, by a teacher who is, all the time, fully aware that science may ultimately prove it to be quite untenable.

His own course went to the opposite extreme. In teaching, where it was possible to let the facts speak for themselves, he did not further urge their bearing upon wider problems. He preferred to warn beginners against drawing superficial inferences in favour of his own general theories, from facts the real meaning of which was not immediately apparent. Father Hahn (S.J.), who studied under him in 1876, writes :—

> One day when I was talking to him, our conversation turned upon evolution. "There is one thing about you I cannot understand," I said, "and I should like a word in explanation. For several months now I have been attending your course, and I have never heard you mention evolution, while in your public lectures everywhere you openly proclaim yourself an evolutionist." [1]
> Now it would be impossible to imagine a better opportunity for insisting on evolution than his lectures on comparative anatomy, when animals are set side by side in respect of the gradual development of functions. But

---

[1] *Revue des Questions Scientifiques* (Brussels), for October 1895.

Huxley was so reserved on this subject in his lectures that, speaking one day of a species forming a transition between two others, he immediately added :—

"When I speak of transition I do not in the least mean to say that one species turned into a second to develop thereafter into a third. What I mean is, that the characters of the second are intermediate between those of the two others. It is as if I were to say that such a Cathedral, Canterbury, for example, is a transition between York Minster and Westminster Abbey. No one would imagine, on hearing the word transition, that a transmutation of these buildings actually took place from one into other."[1]

But to return to his reply :—

"Here in my teaching lectures (he said to me) I have time to put the facts fully before a trained audience. In my public lectures I am obliged to pass rapidly over the facts, and I put forward my personal convictions. And it is for this that people come to hear me."

As to the question whether children should be brought up in entire disregard to the beliefs rejected by himself, but still current among the mass of his fellow-countrymen, he was of opinion that they ought to know "the mythology of their time and country," otherwise one would at the best tend to make young prigs of them ; but as they grew up their questions should be answered frankly.[2]

The natural tendency to veracity, strengthened by the observation of the opposite quality in one with

---

[1] Doubtless in connection with the familiar warning that intermediate types are not necessarily links in the direct line of descent.

[2] The wording of a paragraph in Professor Mivart's "Reminiscences" (*Nineteenth Century*, December 1897, p. 993), tends, I think, to leave a wrong impression on this point.

whom he was early brought into contact, received its decisive impulse, as has been told before, from Carlyle, whose writings confirmed and established his youthful reader in a hatred of shams and make-believes equal to his own.

In his mind no compromise was possible between truth and untruth.[1] Against authorities and influences he published *Man's Place in Nature*, though warned by his friends that to do so meant ruin to his prospects. When he had once led the way and challenged the upholders of conventional orthodoxy, others backed him up with a whole armoury of facts. But his fight was as far as possible for the truth itself, for fact, not merely for controversial victory or personal triumph. Yet, as has been said by a representative of a very different school of thought, who can wonder that he should have hit out straight from the shoulder, in reply to violent or insidious attacks, the stupidity of which sometimes merited scorn as well as anger?

In his theological controversies he was no less careful to avoid any approach to mere abuse or ribaldry such as some opponents of Christian dogma indulged in. For this reason he refused to interpose

---

[1] As he once said, when urged to write a more eulogistic notice of a dead friend than he thought deserved, "The only serious temptations to perjury I have ever known have arisen out of the desire to be of some comfort to people I cared for in trouble. If there are such things as Plato's 'Royal Lies' they are surely those which one is tempted to tell on such occasions. Mrs. —— is such a good devoted little woman, and I am so doubtful about having a soul, that it seems absurd to hesitate to peril it for her satisfaction."

in the well-known Foote case. Discussion, he said, could be carried on effectually without deliberate wounding of others' feelings.

As he wrote in reply to an appeal for help in this case (March 12, 1883) :—

I have not read the writings for which Mr. Foote was prosecuted. But, unless their nature has been grossly misrepresented, I cannot say that I feel disposed to inter-vene on his behalf.

I am ready to go great lengths in defence of freedom of discussion, but I decline to admit that rightful freedom is attacked, when a man is prevented from coarsely and brutally insulting his neighbours' honest beliefs.

I would rather make an effort to get legal penalties inflicted with equal rigour on some of the anti-scientific blasphemers—who are quite as coarse and unmannerly in their attacks on opinions worthy of all respect as Mr. Foote can possibly have been.

The grand result of his determination not to com-promise where truth was concerned, was the securing freedom of thought and speech. One man after another, looking back on his work, declares that if we can say what we think now, it is because he fought the battle of freedom. Not indeed the battle of toleration, if toleration means toleration of error for its own sake. Error, he thought, ought to be extir-pated by all legitimate means, and not assisted because it is conscientiously held.

As Lord Hobhouse wrote, soon after his death :—

I see now many laudatory notices of him in papers. But I have not seen, and I think the younger men do not know, that which (apart from science) I should put forward as his strongest claim to reverence and gratitude ; and that is the steadfast courage and consummate ability with which he fought the battle of intellectual freedom, and insisted that people should be allowed to speak their honest convictions without being oppressed or slandered by the orthodox. He was one of those, perhaps the very foremost, who won that priceless freedom for us ; and, as is too common, people enter into the labours of the brave, and do not even know what their elders endured, or what has been done for themselves.

With this went a proud independence of spirit, intolerant of patronage, careless of titular honours, indifferent to the accumulation of worldly wealth. He cared little even for recognition of his work. "If I had £400 a year,"[1] he exclaimed at the outset of his career, "I should be content to work anonymously for the advancement of science." The only recognition he considered worth having, was that of the scientific world ; yet so little did he seek it, so little insist on questions of priority, that, as Professor Howes tells me, there are at South Kensington among the mass of unpublished drawings from dissections made by him, many which show that he had arrived at discoveries which afterwards brought credit to other investigators.

He was as ready to disclaim for himself any merits which really belonged to his predecessors, whether philosophical or scientific. He was too well read in their works not to be aware of the debt owed them

[1] A sum which might have supported a bachelor, but was entirely inadequate to the needs of a large family.

by his own generation, and he reminded the world
how little the scientific insight of Goethe, for instance,
or the solid labours of Buffon or Reaumur or
Lamarck, deserved oblivion.

The only point on which he did claim recognition
was the honesty of his motives. He was incapable
of doing anything underhand, and he could not bear
even the appearance of such conduct towards his
friends, or those with whom he had business relations.
In such cases he always took the bull by the horns,
acknowledged an oversight or explained what was
capable of misunderstanding. The choice between
Edward Forbes and Hooker for the Royal Society's
medal, or the explanations to Mr. Spencer for not
joining a social reform league of which the latter was
a prominent member, will serve as instances.

The most considerable difference I note among men
(he wrote), is not in their readiness to fall into error, but
in their readiness to acknowledge these inevitable lapses.

For himself, he let no personal feelings stand in the
way when fact negatived his theories : once convinced
that they were untenable, he gave up *Bathybius* and
the European origin of the Horse without hesitation.

The regard in which he was held by his friends
was such that he was sometimes appealed to by both
parties in a dispute. He was a man to be trusted
with the confidence of his friends. " Yes, you are
quite right about 'loyal,' " he writes to Mr. Knowles,
" I love my friends and hate my enemies—which

may not be in accordance with the Gospel, but I have found it a good wearing creed for honest men." But he only regarded as "enemies" those whom he found to be double-dealers, shufflers, insincere, untrustworthy; a fair opponent he respected, and he could agree to differ with a friend without altering his friendship.

A lifelong impression of him was thus summed up by Dr. A. R. Wallace :—

I find that he was my junior by two years, yet he has always seemed to me to be the older, mainly no doubt, because from the very first time I saw him (now more than forty years ago), I recognised his vast superiority in ability, in knowledge, and in all those qualities that enable a man to take a foremost place in the world. I owe him thanks for much kindness and for assistance always cordially given, and although we had many differences of opinion, I never received from him a harsh or unkind word.

To those who could only judge him from his controversial literature, or from a formal business meeting, he often appeared hard and unsympathetic, but never to those who saw beneath the surface. In personal intercourse, if he disliked a man—and a strong individuality has strong likes and dislikes— he would merely veil his feelings under a superabundant politeness of the chilliest kind; but to any one admitted to his friendship he was sympathy itself. And thus, although I have heard him say that his friends, in the fullest sense of the word, could be reckoned on the fingers of one hand, the impression he made upon all who came within the circle of his friendship was such that quite a number felt themselves to possess his intimacy, and one wrote, after his death: "His many private friends are almost tempted to forget the public loss, in thinking of the qualities which so endeared him to them all."

Both the speculative and the practical sides of his intellect were strongly developed. On the one hand, he had an intense love of knowledge, the desire to attain true knowledge of facts, and to organise them in their true relations. His contributions to pure science never fail to illustrate both these tendencies. His earlier researches brought to light new facts in animal life, and new ideas as to the affinities of the creatures he studied; his later investigations were coloured by Darwin's views, and in return contributed no little direct evidence in favour of evolution. But while the progress of the evolution theory in England owed more to his clear and unwearied exposition than to any other cause, while from the first he had indicated the points, such as the causes of sterility and variation, which must be cleared up by further investigation in order to complete the Darwinian theory, he did not add another to the many speculations since put forward.

On the other hand, intense as was his love of pure knowledge, it was balanced by his unceasing desire to apply that knowledge in the guidance of life.

Always feeling that science was not solely for the men of science, but for the people, his constant object was to help the struggling world to ideas which should help them to think truly and so to live rightly. It is still true, he declared, that the people perish for want of knowledge. "If I am to be remembered at all," he writes (see vol. ii. p. 222), "I should like to be remembered as one who did his best to help the people." And again, he says in his Autobiographical Sketch, that other marks of success were as nothing if he could hope that he "had somewhat helped that movement of opinion which has been called the New Reformation."

This kind of aim in his work, of taking up the most fruitful idea of his time and bringing it home to all, is typified by his remark as he entered New York harbour on his visit to America in 1876, and watched the tugs hard at work as they traversed the bay. "If I were not a man," he said, "I think I should like to be a tug."

Two incidents may be cited to show that he did not entirely fail of appreciation among those whom he tried to help. Speaking of the year 1874, Professor Mivart writes ("Reminiscences of T. H. Huxley," *Nineteenth Century*, Dec. 1897) :—

I recollect going with him and Mr. John Westlake, Q.C., to a meeting of artisans in the Blackfriars Road, to whom he gave a friendly address. He felt a strong interest in working-men, and was much beloved by them. On one occasion, having taken a cab home, on his arrival there, when he held out his fare to the cabman, the latter replied, "Oh no, Professor, I have had too much pleasure and profit from hearing you lecture to take any money from your pocket—proud to have driven you, sir !

The other is from a letter to the *Pall Mall Gazette* of September 20, 1892, from Mr. Raymond Blaythwayt, on "The Uses of Sentiment":—

Only to-day I had a most striking instance of sentiment come beneath my notice   I was about to enter my house, when a plain, simply-dressed working-man came up to me with a note in his hand, and touching his hat, he said, "I think this is for you, sir," and then he added, "Will you give me the envelope, sir, as a great favour?" I looked at it, and seeing it bore the signature of Professor Huxley, I replied, "Certainly I will; but why do you ask for it?" "Well," said he, "it's got Professor Huxley's signature, and it will be something for me to show my mates and keep for my children. He have done me and my like a lot of good; no man more."

In practical administration, his judgment of men, his rapid perception of the essential points at issue, his observance of the necessary limits of official forms, combined with the greatest possible elasticity within these limits, made him extremely successful.

As Professor (writes the late Professor Jeffery Parker), Huxley's rule was characterised by what is undoubtedly the best policy for the head of a department.   To a new subordinate, "The General," as he was always called, was rather stern and exacting, but when once he was convinced that his man was to be trusted, he practically let him take his own course; never interfered in matters of detail, accepted suggestions with the greatest courtesy and good humour, and was always ready with a kindly and humorous word of encouragement in times of difficulty. I was once grumbling to him about how hard it was to carry on the work of the laboratory through a long series of November fogs, "when neither sun nor stars in many days appeared." "Never mind, Parker," he said, instantly capping my quotation, "cast four anchors out of the stern and wish for day."

Nothing, indeed, better illustrates this willingness to listen to suggested improvements than the inversion of the order of studies in the biological course which he inaugurated in 1872, namely, the substitution of the anatomy of a vertebrate for the microscopic examination of a unicellular organism as the opening study. This was entirely Parker's doing. "As one privileged at the time to play a minor part," writes Professor Howes (*Nature*, January 6, 1898, p. 228), "I well recall the determination in Parker's mind that the change was desirable, and in Huxley's, that it was not. Again and again did Parker appeal in vain, until at last, on the morning of October 2, 1878, he triumphed."

On his students he made a deep and lasting impression.

His lectures (writes Jeffery Parker) were like his writings, luminously clear, without the faintest disposition to descend to the level of his audience; eloquent, but with no trace of the empty rhetoric which so often does duty for that quality; full of a high seriousness, but with no suspicion of pedantry; lightened by an occasional epigram or flashes of caustic humour, but with none of the small jocularity in which it is such a temptation to a lecturer to indulge. As one listened to him one felt that comparative anatomy was indeed worthy of the devotion of a life, and that to solve a morphological problem was as fine a thing as to win a battle. He was an admirable draughtsman, and his blackboard illustrations were always a great feature of his lectures, especially when, to show the relation of two animal types, he would, by a few rapid strokes and smudges, evolve the one into the other before our eyes. He seemed to have a real affection for some of the specimens illustrating his lectures, and would handle them in a peculiarly loving manner; when he was lecturing on man, for instance, he would sometimes throw his arm over the shoulder of the skeleton beside him and take its hand, as if its silent companionship were an inspiration. To me his lectures before his small class at Jermyn Street or South Kensington were almost more impressive than the discourses at the Royal Institution, where for an hour and a half he poured forth a stream of dignified, earnest, sincere words in perfect literary form, and without the assistance of a note.

Another description is from the pen of an old pupil in the autumn of 1876, Professor H. Fairfield Osborn, of Columbia College :—

Huxley, as a teacher, can never be forgotten by any of his students. He entered the lecture-room promptly as the clock was striking nine,[1] rather quickly, and with his head bent forward "as if oppressive with its mind." He usually glanced attention to his class of about ninety, and began speaking before he reached his chair. He spoke between his lips, but with perfectly clear analysis,

---

[1] In most years the lectures began at ten.

with thorough interest, and with philosophic insight which was far above the average of his students. He used very few charts, but handled the chalk with great skill, sketching out the anatomy of an animal as if it were a transparent object. As in Darwin's face, and as in Erasmus Darwin's or Buffon's, and many other anatomists with a strong sense of form, his eyes were heavily overhung by a projecting forehead and eyebrows, and seemed at times to look inward. His lips were firm and closely set, with the expression of positiveness, and the other feature which most marked him was the very heavy mass of hair falling over his forehead, which he would frequently stroke or toss back. Occasionally he would light up the monotony of anatomical description by a bit of humour.

Huxley was the father of modern laboratory instruction; but in 1879 he was so intensely engrossed with his own researches that he very seldom came through the laboratory, which was ably directed by T. Jeffery Parker, assisted by Howes and W. Newton Parker, all of whom are now professors, Howes having succeeded to Huxley's chair. Each visit, therefore, inspired a certain amount of terror, which was really unwarranted, for Huxley always spoke in the kindest tones to his students, although sometimes he could not resist making fun at their expense. There was an Irish student who sat in front of me, whose anatomical drawings in water-colour were certainly most remarkable productions. Huxley, in turning over his drawing-book, paused at a large blur, under which was carefully inscribed, " sheep's liver," and smilingly said, " I am glad to know that is a liver; it reminds me as much of Cologne cathedral in a fog as of anything I have ever seen before." Fortunately the nationality of the student enabled him to fully appreciate the humour.

The same note is sounded in Professor Mivart's description of these lectures in his Reminiscences :—

The great value of Huxley's anatomical ideas, and the admirable clearness with which he explained them, led me in the autumn of 1861 to seek admission as a student to his course of lectures at the School of Mines in Jermyn Street. When I entered his small room there to make this request, he was giving the finishing touches to a dissection of part of the nervous system of a skate, worked out for the benefit of his students. He welcomed my application with the greatest cordiality, save that he insisted I should be only an honorary student, or rather, should assist at his lectures as a friend. I availed myself of his permission on the very next day, and subsequently attended almost all his lectures there and elsewhere, so that he one day said to me, "I shall call you my 'constant reader.'" To be such a reader was to me an inestimable privilege, and so I shall ever consider it. I have heard many men lecture, but I never heard any one lecture as did Professor Huxley. He was my very ideal of a lecturer. Distinct in utterance, with an agreeable voice, lucid as it was possible to be in exposition, with admirably chosen language, sufficiently rapid, yet never hurried, often impressive in manner, yet never otherwise than completely natural, and sometimes allowing his audience a glimpse of that rich fund of humour ever ready to well forth when occasion permitted, sometimes accompanied with an extra gleam in his bright dark eyes, sometimes expressed with a dryness and gravity of look which gave it a double zest.

I shall never forget the first time I saw him enter his lecture-room. He came in rapidly, yet without bustle, and as the clock struck, a brief glance at his audience and then at once to work. He had the excellent habit of beginning each lecture (save, of course, the first) with a recapitulation of the main points of the preceding one. The course was amply illustrated by excellent coloured diagrams, which, I believe, he had made; but still more valuable were the chalk sketches he would draw on the blackboard with admirable facility, while he was talking, his rapid, dexterous strokes quickly building up an organism in our minds, simultaneously through ear and eye. The lecture over, he was ever ready to answer questions, and I often admired his patience in explaining points which there was no excuse for any one not having understood.

Still more was I struck with the great pleasure which he showed when he saw that some special points of his teaching had not only been comprehended, but had borne fruit, by their suggestiveness in an appreciative mind.

To one point I desire specially to bear witness. There were persons who dreaded sending young men to him, fearing lest their young friends' religious beliefs should be upset by what they might hear said. For years I attended his lectures, but never once did I hear him make use of his position as a teacher to inculcate, or even hint at, his own theological views, or to depreciate or assail what might be supposed to be the religion of his hearers. No one could have behaved more loyally in that respect, and a proof that I thought so is that I subsequently sent my own son to be his pupil at South Kensington, where his experience confirmed what had previously been my own.

As to science, I learnt more from him in two years than I had acquired in any previous decade of biological study.

The picture is completed by Professor Howes in the *Students' Magazine* of the Royal College of Science :—

As a class lecturer Huxley was *facile princeps*, and only those who were privileged to sit under him can form a conception of his delivery. Clear, deliberate, never hesitant nor unduly emphatic, never repetitional, always logical, his every word told. Great, however, as were his class lectures, his working-men's were greater. Huxley was a firm believer in the "distillatio per ascensum" of scientific knowledge and culture, and spared no pains in approaching the artisan and so-called "working classes." He gave the workmen of his best. The substance of his *Man's Place in Nature*, one of the most successful and popular of his writings, and of his *Crayfish*, perhaps the most perfect zoological treatise ever published, was first communicated to them. In one of the last conversations I had with him, I asked his views on the desirability of discontinuing the workmen's lectures at Jermyn Street, since the development of working-men's colleges and institutes is regarded by some to have rendered their continuance unnecessary. He replied, almost with indignation, "With our central situation and resources, we ought to be in a position to give the workmen that which they cannot get elsewhere," adding that he would deeply deplore any such discontinuance.

And now, a word or two concerning Huxley's personal conduct towards his pupils, hearers, and subordinates.

As an examiner he was most just, aiming only to ascertain the examinee's knowledge of fundamentals, his powers of work, and the manner in which he had been taught. A country school lad came near the boundary line in the examination; though generally weak, his worst fault was a confusion of the parts of the heart. In his description of that organ he had transposed the valves. On appeal, Huxley let him through, observing, most characteristically, "Poor little beggar, I never got them correctly myself until I reflected that a bishop was never in the right." [1] Again, a student of more advanced years, of the "mugging" type, who had come off with flying colours in an elementary examination, showed signs of uneasiness as the advanced one approached. "Stick an observation into him," said

---

[1] The "mitral" valve being on the left side.

Huxley. It was stuck, and acted like a stiletto, a jump into the air and utter collapse being the result.

With his hearers Huxley was most sympathetic. He always assumed absolute ignorance on their part, and took nothing for granted.[1] When time permitted, he would remain after a lecture to answer questions; and in connection with his so doing his wonderful power of gauging and rising to a situation, once came out most forcibly. Turning to a student, he asked, "Well, I hope you understood it all." "All, sir, but one part, during which you stood between me and the blackboard," was the reply: the rejoinder, "I did my best to make myself clear, but could not render myself transparent." Quick of comprehension and of action, he would stand no nonsense. The would-be teacher who, wholly un-fitted by nature for educational work, was momentarily dismissed, realised this, let us hope to his advantage. And the man suspected of taking notes of Huxley's lectures for publication unauthorised, probably learned the lesson of his life, on being reminded that, in the first place, a lecture was the property of the person who delivered it, and, in the second, he was not the first person who had mistaken aspiration for inspiration.

Though candid, Huxley was never unkind. . . .

Huxley never forgot a kindly action, never forsook a friend, nor allowed a labour to go unrewarded. In testimony to his sympathy to those about him and his self-sacrifice for the cause of science, it may be stated that in the old days, when the professors took the fees and disbursed the working expenses of the laboratories, he, doing this at a loss, would refund the fees of students whose position, from friendship or special circumstances, was exceptional.

As for his lectures and addresses to the public,

---

[1] This was a maxim on lecturing, adopted from Faraday.

they used to be thronged by crowds of attentive listeners.

Huxley's public addresses (writes Professor Osborn) always gave me the impression of being largely impromptu; but he once told me: "I always think out carefully every word I am going to say. There is no greater danger than the so-called *inspiration of the moment*, which leads you to say something which is not exactly true, or which you would regret afterwards."

Mr. G. W. Smalley has also left a striking description of him as a lecturer in the seventies and early eighties.

I used always to admire the simple and business-like way in which Huxley made his entry on great occasions. He hated anything like display, and would have none of it. At the Royal Institution, more than almost anywhere else, the lecturer, on whom the concentric circles of spectators in their steep amphitheatre look down, focuses the gaze. Huxley never seemed aware that anybody was looking at him. From self-consciousness he was, here as elsewhere, singularly free, as from self-assertion. He walked in through the door on the left, as if he were entering his own laboratory. In these days he bore scarcely a mark of age. He was in the full vigour of manhood and looked the man he was. Faultlessly dressed—the rule in the Royal Institution is evening costume—with a firm step and easy bearing, he took his place apparently without a thought of the people who were cheering him. To him it was an anniversary. He looked, and he probably was, the master. Surrounded as he was by the celebrities of science and the ornaments of London drawing-rooms, there was none who had quite the same kind of intellectual ascendency which belonged to him. The square forehead, the square jaw, the tense lines of the mouth, the deep flashing dark eyes, the impression of something more than strength he gave you, an impression of sincerity, of solid force, of immovability, yet with the gentleness arising from the serene consciousness of his strength—all this belonged to Huxley and to him alone. The first glance magnetised his audience. The eyes were those of one accustomed to command, of one having authority, and not fearing on occasion to use it. The hair swept carelessly away from the broad forehead and grew rather long behind, yet the length did not suggest, as it often does, effeminacy. He was masculine in everything—look, gesture, speech. Sparing of gesture, sparing of emphasis, careless of mere rhetorical or oratorical art, he had nevertheless the secret of the highest art of all, whether in oratory or whatever else— he had simplicity. The force was in the thought and the diction, and he needed no other. The voice was rather deep, low, but quite audible, at times sonorous, and always full. He used the chest-notes. His manner here, in the presence of this select and rather limited audience—for the theatre of the Royal Institution holds, I think, less than a thousand people—was exactly the same as before a great company whom he addressed at [Liverpool], as President of the British Association for the Advancement of Science. I remember going late to that, and having to sit far back, yet hearing every word easily ; and there too the feeling was the same, that he had mastered his audience, taken possession of them, and held them to the end in an unrelaxing grip, as a great actor at his best does. There was nothing of the actor about him, except that he knew how to stand still, but masterful he ever was.

Up to the time of his last illness, he regularly breakfasted at eight, and avoided, as far as possible, going out to that meal, a "detestable habit" as he called it, which put him off for the whole day. He left the house about nine, and from that time till midnight at earliest was incessantly busy. His regular lectures involved an immensity of labour, for he would never make a statement in them which he had not personally verified by experiment. In the Jermyn Street days he habitually made preparations to illustrate the points on which he was lecturing, for his students had no laboratory in which to work out the things for themselves. His lectures to working-men also involved as much careful preparation as the more conspicuous discourses at the Royal Institution.

This thoroughness of preparation had no less effect on the teacher than on the taught. He writes to an old pupil :—

It is pleasant when the "bread cast upon the water" returns after many days; and if the crumbs given in my lectures have had anything to do with the success on which I congratulate you, I am very glad.

I used to say of my own lectures that if nobody else learned anything from them, I did ; because I always took a great deal of pains over them. But it is none the less satisfactory to find that there *were* other learners.

As for the ordinary course of a day's work, the more fitful energy and useless mornings of the earliest period in London were soon left behind. He was never one of those portentously early risers who do a fair day's work before other people are up ; there was only one period, about 1873, when he had to be specially careful of his health, and, under Sir Andrew Clark's regime, took riding exercise for an hour each day before starting for South Kensington, that he records the fact of doing any work before breakfast, and that was letter-writing.

Much of the day during the session, and still more when his lectures were over, would thus be spent in original research, or in the examination and description of fossils in his official duty as Paleontologist to the Survey. As often as not, there would be a sitting of some Royal Commission to attend; committees of some learned society; meetings or dinners in the evening; if not, there would be an article to write or proofs to correct. Indeed, the greater part of the work by which the world knows him best was done after dinner, and after a long day's work in the lecture-room and laboratory.

He possessed a wonderful faculty for tearing out the heart of a book, reading it through at a gallop, but knowing what it said on all the points that interested him. Of verbal memory he had very little; in spite of all his reading I do not believe he knew half a dozen consecutive lines of poetry by heart. What he did know was the substance of what an author had written; how it fitted into his own scheme of knowledge; and where to find any point again when he wished to cite it.

In his biological studies his immense knowledge was firmly fixed in his mind by practical investigation; as is said above, he would take at second hand

nothing for which he vouched in his teaching, and
was always ready to repeat for himself the experiments of others, which determined questions of
interest to him. The citations, analyses, maps, with
which he frequently accompanied his reading, were
all part of the same method of acquiring facts and
setting them in order within his mind. So careful,
indeed, was he in giving nothing at second hand,
that one of his scientific friends reproached him
with wasting his time upon unnecessary scientific
work, to which competent investigators had already
given the stamp of their authority. "Poor——,"
was his comment afterwards, "if that is his own
practice, his work will never live." On the literary
side, he was omnivorous—consuming everything, as
Mr. Spencer put it, from fairy tales to the last volume
on metaphysics.

Unlike Darwin, to whom scientific research was
at length the only thing engrossing enough to make
him oblivious of his never-ending ill-health, to the
gradual exclusion of other interests, literary and
artistic, Huxley never lost his delight in literature or
in art. He had a keen eye for a picture or a piece
of sculpture, for, in addition to the draughtsman's and
anatomist's sense of form, he had a strong sense of
colour. To good music he was always susceptible.[1]
He played no instrument ; as a young man, however,

[1] To one breaking in upon him at certain afternoon hours in
his room at South Kensington, "a whiff of the pipe" (writes
Professor Howes), "and a snatch of some choice melody or a
Bach's fugue, were the not infrequent welcome."

he used to sing a little, but his voice, though true, was never strong. But he had small leisure to devote to art. On his holidays he would sometimes sketch with a firm and rapid touch. His illustrations to the *Cruise of the Rattlesnake* show what his untrained capacities were. But to go to a concert or opera was rare after middle life; to go to the theatre rarer still, much as he appreciated a good play His time was too deeply mortgaged; and in later life, the deafness which grew upon him added a new difficulty.

In poetry he was sensitive both to matter and form. One school of modern poetry he dismissed as "sensuous caterwauling": a busy man, time and patience failed him to wade through the trivial discursiveness of so much of Wordsworth's verse; thus unfortunately he never realised the full value of a poet in whom the mass of ore bears so large a proportion to the pure metal. Shelley was too diffuse to be among his first favourites; but for simple beauty, Keats; for that, and for the comprehension of the meaning of modern science, Tennyson; for strength and feeling, Browning as represented by his earlier poems—these were the favourites among the moderns. He knew his eighteenth-century classics, but knew better his Milton and his Shakespeare, to whom he turned with ever-increasing satisfaction, as men do who have lived a full life.

His early acquaintance with German had given him a lasting admiration of the greatest representatives of German literature, Goethe above all, in whose writings he found a moral grandeur to be ranked with that of the Hebrew prophets. Eager to read Dante in the original, he spent much of his leisure on board the *Rattlesnake* in making out the Italian with the aid of a dictionary, and in this way came to know the beauties of the *Divina Commedia*. On the other hand, it was a scientific interest which led him in later life to take up his Greek, though one use he put it to was to read Homer in the original.

Though he was a great novel-reader, and, as he grew older, would always have a novel ready to take up for a while in the evening, his chief reading, in German and French as well as English, was philosophy and history.

His recreations were, as a rule, literary, and consisted in a change of mental occupation. The only times I can remember his playing an outdoor game are in the late sixties, when he started his elder children at cricket on the common at Littlehampton, and in 1871 when he played golf at St. Andrews. When first married, he promised his wife to reserve Saturday afternoons for recreation, and constantly went with her to the Ella concerts. About 1861 she urged him to take exercise by joining Mr. Herbert Spencer at racquets ; but the pressure of work before long absorbed all his time. In his youth he was extremely fond of chess, and played eagerly with his fellow-students at Charing Cross Hospital or with his messmates on board the *Rattlesnake*. But after he taught me the game, somewhere about 1869 or 1870, I do not think he ever found time for it again.

His principal exercise was walking during the holidays. In his earlier days especially, when over-wrought by the stress of his life in London, he used to go off with a friend for a week's walking tour in Wales or the Lakes, in Brittany or the Eifel country, or in summer for a longer trip to Switzerland. In in this way he " burnt up the waste products," as he would say, of his town life, and came back fresh for a new spell of unintermittent work.

But, on the whole, the amount of exercise he took was insufficient for his bodily needs. Even the riding prescribed for him when he first broke down, became irksome, and was not continued very long, although his bodily machine was such as could only be kept in perfect working order by more exercise than he would give. His physique was not adapted to burn up the waste without special stimulus. I remember once, as he and I were walking up Beachy Head, we passed a man with a splendid big chest. " Ah," said my father regretfully, " if I had only had a chest like that, what a lot of work I could have done."

When, in 1872, he built his new house in Marl-borough Place, my father bargained for two points; one, that each member of the family should have a corner of his or her own, where, as he used to say, it would be possible to " consume their own smoke "; the other, that the common living-rooms should be of

ample size. Thus from 1874 onwards he was enabled
to see something of his many friends who would come
as far as St. John's Wood on a Sunday evening. No
formal invitation for a special day was needed. The
guests came, before supper or after, sometimes more,
sometimes fewer, as on any ordinary at-home day.
There was a simple informal meal at 6.30 or 7 o'clock,
which called itself by no more dignified name than
high tea—was, in fact, a cold supper with varying
possibilities in the direction of dinner or tea. It was
a chance medley of old and young—friends of the
parents and friends of the children, but all ultimately
centring round the host himself, whose end of the
table never flagged for conversation, grave or gay.

Afterwards talk would go on in the drawing-room,
or, on warm summer evenings, in the garden—nothing
very extensive, but boasting a lawn with an old apple-
tree at the further end, and in the borders such flowers
and trees as endure London air. Later on, there was
almost sure to be some music, to which my father
himself was devoted. His daughters sang ; a musical
friend would be there ; Mr. Herbert Spencer, a
frequent visitor, was an authority on music. Once
only do I recollect any other form of entertainment,
and that was an occasion when Sir Henry Irving, then
not long established at the Lyceum, was present and
recited " Eugene Aram " with great effect.

In his *London Letters* Mr. G. W. Smalley [1] has

[1] Another interesting account from the same pen is to be found
in the article " Mr. Huxley," *Scribner's Magazine*, October 1895.

recorded his impressions of these evenings, at which
he was often present :—

> There used to be Sunday evening dinners and parties
> in Marlborough Place, to which people from many other
> worlds than those of abstract science were bidden ; where
> talk was to be heard of a kind rare in any world.   It
> was scientific at times, but subdued to the necessities of
> the occasion ; speculative, yet kept within such bounds
> that bishop or archbishop might have listened without
> offence ; political even, and still not commonplace ; literary
> without pretence, and when artistic, free from affectation.
>   There and elsewhere Mr. Huxley easily took the lead
> if he cared to, or if challenged.   Nobody was more ready
> in a greater variety of topics, and if they were scientific
> it was almost always another who introduced them.
> Unlike some of his comrades of the Royal Society, he
> was of opinion that man does not live by science alone,
> and nothing came amiss to him.   All his life long he has
> been in the front of the battle that has raged between
> science and—not religion, but theology in its more dog-
> matic form.   Even in private the alarm of war is some-
> times heard, and Mr. Huxley is not a whit less formidable
> as a disputant across the table than with pen in hand.
> Yet an angry man must be very angry indeed before he
> could be angry with this adversary.   He disarmed his
> enemies with an amiable grace that made defeat endurable
> if not entirely delightful.

As for his method of handling scientific subjects in
conversation :—

> He has the same quality, the same luminous style of
> exposition, with which his printed books have made all
> readers in America and England familiar.   Yet it has
> more than that.   You cannot listen to him without
> thinking more of the speaker than of his science,

more of the solid beautiful nature than of the intellectual gifts, more of his manly simplicity and sincerity than of all his knowledge and his long services.

But his personality left the deepest impression, perhaps, upon those who studied under him and worked with him longest, before taking their place elsewhere in the front ranks of biological science.

With him (Professor A. Hubrecht[1] writes), we his younger disciples, always felt that in acute criticism and vast learning nobody surpassed him, but still what we yet more admired than his learning was his wisdom. It was always a delight to read any new article or essay from his pen, but it was an ever so much higher delight to hear him talk for five minutes. His was the most beautiful and the most manly intellect I ever knew of.

So, too, Professor E. Ray Lankester :—

There has been no man or woman whom I have met on my journey through life, whom I have loved and regarded as I have him, and I feel that the world has shrunk and become a poor thing, now that his splendid spirit and delightful presence are gone from it. Ever since I was a little boy he has been my ideal and hero.

While the late Jeffery Parker concludes his Recollections with these words :—

Whether a professor is usually a hero to his demonstrator I cannot say ; I only know that, looking back across an interval of many years and a distance of half the circumference of the globe, I have never ceased to be impressed with the manliness and sincerity of his character, his complete honesty of purpose, his high moral standard,

---

[1] Of Utrecht University.

his scorn of everything mean or shifty, his firm determination to speak what he held to be truth at whatever cost of popularity. And for these things "I loved the man, and do honour to his memory, on this side idolatry, as much as any."

Even those who scarcely knew him apart from his books, underwent the influence of that "determination to speak what he held to be truth." I may perhaps be allowed to quote in illustration two passages from letters to myself—one written by a woman, the other by a man :—

" ' The surest-footed guide' is exactly true, to my feeling. Everybody else, among the great, used to disappoint one somewhere. He—never ! "

" He was so splendidly brave that one can never repay one's debt to him for his example. He made all pretence about religious belief, and the kind of half-thinking things out, and putting up in a slovenly way with half-formed conclusions, seem the base thing which it really is."

CHAPTER XVI

1895

I HAVE often regretted that I did not regularly take notes of my father's conversation, which was striking, not so much for the manner of it—though that was at once copious and crisp,—as for the strength and substance of what he said. Yet the striking fact, the bit of philosophy, the closely knitted argument, were perfectly unstudied, and as in other most interesting talkers, dropped into the flow of conversation as naturally as would the more ordinary experiences of less richly stored minds.

However, in January 1895 I was staying at Eastbourne, and jotted down several fragments of talk as nearly as I could recollect them. Conversation not immediately noted down I hardly dare venture upon, save perhaps such an unforgettable phrase as this, which I remember his using one day as we walked on the hills near Great Hampden :—"It is one of the most saddening things in life that, try as we may, we can never be certain of making people happy, whereas we can almost always be certain of making them unhappy."

January 16.—At lunch he spoke of Dr. Louis Robinson's experiments upon simian characteristics in new-born children.   He himself had called attention before to the incurved feet of infants, but the power of hanging by the hands was a new and important discovery.[1]

He expressed his disgust with a certain member of the Psychical Research Society for his attitude towards spiritualism : " He doesn't believe in it, yet lends it the cover of his name.   He is one of the people who talk of the 'possibility' of the thing, who think the difficulties of disproving a thing as good as direct evidence in its favour."

He thought it hard to be attacked for " the contempt of the man of science " when he was dragged into debate by Mr. Andrew Lang's *Cock Lane and Common Sense*, he saying in a very polite letter : " I am content to leave Mr. Lang the Cock Lane Ghost if I may keep common sense."   " After all," he added, " when a man has been through life and made his judgments, he must have come to a decision that there are some subjects it is not worth while going into."

January 18.—I referred to an article in the last *Nineteenth Century*, and he said :—" As soon as I saw it, I wrote, 'Knowles, my friend, you don't draw me

---

[1] Professor H. F. Osborn tells this story of his :—" When a fond mother calls upon me to admire her baby, I never fail to respond ; and while cooing appropriately, I take advantage of an opportunity to gently ascertain whether the soles of its feet turn in, and tend to support my theory of arboreal descent."

this time. If a man goes on attributing statements to me which I have shown over and over again—giving chapter and verse—to be the contrary of what I did say, it is no good saying any more.'"

But would not this course of silence leave the mass of the British public believing the statements of the writer?

"The mass of the public will believe in ten years precisely the opposite of what they believe now. If a man is not a fool, it does him no harm to be believed one. If he really is a fool, it does matter. There never was book so derided and scoffed at as my first book, *Man's Place in Nature*, but it was true, and I don't know I was any the worse for the ridicule.

"People call me fond of controversy, but, as a fact, for the last twenty years at all events, I have never entered upon a controversy without some further purpose in view. As to Gladstone and his *Impregnable Rock*, it wasn't worth attacking them for themselves; but it was most important at that moment to shake him in the minds of sensible men.

"The movement of modern philosophy is back towards the position of the old Ionian philosophers, but strengthened and clarified by sound scientific ideas. If I publish my criticism on Comte, I should have to re-write it as a summary of philosophical ideas from the earliest times. The thread of philosophical development is not on the lines usually laid down for it. It goes from Democritus and the rest to the Epicureans, and then the Stoics, who tried to reconcile it with popular theological ideas, just as was done by the Christian Fathers. In the Middle Ages it was entirely lost under the theological theories of the time; but reappeared with Spinoza, who, however, muddled it up with a lot of metaphysics which made him almost unintelligible.

"Plato was the founder of all the vague and unsound thinking that has burdened philosophy, deserting facts for possibilities, and then, after long and beautiful stories of what might be, telling you he doesn't quite believe them himself.

"A certain time since it was heresy to breathe a word against Plato; but I have a nice story of Sir Henry Holland. He used to have all the rising young men to breakfast, and turn out their latest ideas. One morning I went to breakfast with him, and we got into very intimate conversation, when he wound up by saying, 'In my opinion, Plato was an ass! But don't tell any one I said so.'"

We talked on geographical teaching; he began by insisting on the need of a map of the earth (on the true scale) showing the insignificance of all elevations and depressions on the surface. Secondly, one should take any place as centre, and draw about it circles of 50 or 100 miles radius, and see what lies within them; and note the extent of the influence exerted by the central point. At the same time, one should always compare the British Isles to scale. For instance, the Ægean is about as big as Britain; while the smallness of Judæa is remarkable. After the Exile, the Jewish part was about as big as the county of Gloucester.   How few boys realise this, though they are taught classical geography.

"The real chosen people were the Greeks.   One of the most remarkable things about them is not only the smallness, but the late rise of Attica, whereas Magna Græcia flourished in the eighth century. The Greeks were doing everything—piracy, trade, fighting, expelling the Persians.   Never was there so large a number of self-governing communities.

"They fell short of the Jews in morality.   How curious is the tolerant attitude of Socrates, like a modern man of the world talking to a young fellow who runs after the girls.   The Jew, however he fell short in other respects, set himself a certain standard in cleanliness of life, and would not fall below it. The more creditable to him, because these vices were the offspring of the Semitic races among whom the Jew lived.

"There is a curious similarity between the position of the Jew in ancient times and what it is now.   They were procurers and usurers among the Gentiles, yet many of them were singularly high-minded and pure. All too with an intense clannishness, the secret of their success, and a sense of superiority to the Gentile which would prevent the meanest Jew from sitting at table with a proconsul.

"The most remarkable achievement of the Jew was to impose on Europe for eighteen centuries his own superstitions—his ideas of the supernatural.

Jahveh was no more than Zeus or Milcom; yet the Jew got established the belief in the inspiration of his Bible and his Law. If I were a Jew, I should have the same contempt as he has for the Christian who acted in this way towards me, who took my ideas and scorned me for clinging to them."

January 21.—Yesterday evening he again declared that it was very hard for a man of peace like himself to have been dragged into so many controversies. "I declare that for the last twenty years I have never attacked, but always fought in self-defence, counting Darwin, of course, as part of myself, for dear Darwin never could nor would defend himself. Before that, I admit I attacked ——, but I could not trust the man." A pause. "No, there was one other case, when I attacked without being directly assailed, and that was Gladstone. But it was good for other reasons. It has always astonished me how a man after fifty or sixty years of life among men could be so ignorant of the best way to handle his materials. If he had only read Dana, he would have found his case much better stated than ever he stated it. He seemed never to have read the leading authorities on his own side."

Speaking of the hesitation shown by the Senate of London University in grappling with a threatened obstacle to reform, he remarked: "It is very strange how most men will do anything to evade responsibility."

January 23.—At dinner the talk turned on plays. Mr. H. A. Jones had sent him *Judah*, which he thought good, though "there must be some hostility—except in the very greatest writers—between the dramatic and the literary faculties. I noticed many points I objected to, but felt sure they met with applause. Indeed in the theatre I have noticed that what I thought the worst blots on a piece invariably brought down the house."

He remarked how the French, in dramatic just as in artistic matters, are so much better than the English in composition, in avoiding anything slipshod in the details, though the English artists draw just as well and colour perhaps better.

The following sketch of human character is not actually a fragment of conversation, though it might almost pass for such ; it comes from a letter to Mrs. W. K. Clifford, of February 10, 1895 :—

Men, my dear, are very queer animals, a mixture of horse-nervousness, ass-stubbornness and camel-malice— with an angel bobbing about unexpectedly like the apple in the posset, and when they can do exactly as they please, they are very hard to drive.

Whatever he talked of, his talk never failed to impress those who conversed with him. One or two such impressions have been recorded. Mr. Wilfrid Ward, whose interests lie chiefly in philosophy and theology, was his neighbour at Eastbourne, and in the *Nineteenth Century* for August 1896 has given various reminiscences of their friendly intercourse.

His conversation (he writes) was singularly finished, and (if I may so express it) clean cut ; never long-winded or prosy; enlivened by vivid illustrations. He was an excellent *raconteur*, and his stories had a stamp of their own which would have made them always and everywhere acceptable. His sense of humour and economy of words would have made it impossible, had he lived to ninety, that they should ever have been disparaged as symptoms of what has been called "anecdotage."

One drawback to conversation, however, he began to complain of during the later seventies.

It is a great misfortune (he remarked to Professor Osborn) to be deaf in only one ear. Every time I dine out the lady sitting by my good ear thinks I am charming, but I make a mortal enemy of the lady on my deaf side.

In ordinary conversation he never plunged at once into deep subjects. His welcome to the newcomer was always of the simplest and most unstudied. He had no mannerisms nor affectation of phrase. He would begin at once to talk on everyday topics; an intimate friend he would perhaps rally upon some standing subject of persiflage. But the subsequent course of conversation adapted itself to his company. Deeper subjects were reached soon enough by those who cared for them; with others he was quite happy to talk of politics or people or his garden, yet, whatever he touched, never failing to infuse into it an unexpected interest.

In this connection, a typical story was told me by a great friend of mine, whom we had come to know through his marriage with an early friend of the family. "Going to call at Hodeslea," he said, "I was in some trepidation, because I didn't know anything about science or philosophy ; but when your mother began to talk over old times with my wife, your father came across the room and sat down by me, and began to talk about the dog which we had brought with us. From that he got on to the different races of dogs and their origin and connections, all quite simply, and not as though to give information, but just to talk about something which obviously interested me   I shall never forget how extraordinarily kind it was of your father to take all this trouble in entertaining a complete stranger, and choosing a subject which put me at my ease at once, while he told me all manner of new and interesting things."

A few more fragments of his conversation have been preserved—the following by Mr. Wilfrid Ward. Speaking of Tennyson's conversation, he said :—

Doric beauty is its characteristic—perfect simplicity, without any ornament or anything artificial.

Telling how he had been to a meeting of the British Museum Trustees, he said :—

After the meeting, Archbishop Benson helped me on with my great-coat. I was *quite overcome* by this species of spiritual investiture. "Thank you, Archbishop," I said, "I feel as if I were receiving the *pallium.*"

Speaking of two men of letters, with neither of whom he sympathised, he once said :—

Don't mistake me. One is a thinker and man of letters, the other is only a literary man. Erasmus was a man of letters, Gigadibs a literary man. A.B. is the incarnation of Gigadibs. I should call him *Gigadibsius Optimus Maximus.*

Another time, referring to Dean Stanley's historical impressionability, as militating against his sympathies with Colenso, he said :—

Stanley could believe in anything of which he had seen the supposed site, but was sceptical where he had not seen. At a breakfast at Monckton Milnes's, just at the time of the Colenso row, Milnes asked me my views on the Pentateuch, and I gave them. Stanley differed from me. The account of Creation in Genesis he dismissed at once as unhistorical; but the call of Abraham, and the historical narrative of the Pentateuch, he accepted. This was because he had seen Palestine— but he wasn't present at the Creation.

When he and Stanley met, there was sure to be a brisk interchange of repartee. One of these occasions, a ballot day at the Athenæum, has been recorded by the late Sir W. H. Flower :—

A well-known popular preacher of the Scotch Presbyterian Church, who had made himself famous by predictions of the speedy coming of the end of the world, was up for election. I was standing by Huxley when the Dean, coming straight from the ballot boxes, turned towards us. "Well," said Huxley, "have you been voting for C. ?" "Yes, indeed I have," replied the Dean. "Oh, I thought the priests were always opposed to the prophets," said Huxley. "Ah !" replied the Dean, with that well-known twinkle in his eye, and the sweetest of smiles, "but you see, I do not believe in his prophecies, and some people say I am not much of a priest."

A few words as to his home life may perhaps be fitly introduced here. Towards his children he had the same union of underlying tenderness veiled beneath inflexible determination for what was right, which marked his intercourse with those outside his family.

As children we were fully conscious of this side of his character. We felt our little hypocrisies shrivel up before him; we felt a confidence in the infallible rectitude of his moral judgments which inspired a kind of awe. His arbitrament was instant and final, though rarely invoked, and was perhaps the more tremendous in proportion to its rarity. This aspect, as if of an oracle without appeal, was heightened in our minds by the fact that we saw but little of him. This was one of the penalties of his hard-driven existence. In the struggle to keep his head above water for the first fifteen or twenty years of his married life, he had scarcely any time to devote to his children. The "lodger," as he used to call himself at one time, who went out early and came back late, could sometimes spare half an hour just before or after dinner to draw wonderful pictures for the little ones, and these were memorable occasions. I remember that he used to profess a horror of being too closely watched, or of receiving suggestions, while he drew. "Take care, take care," he would exclaim, "or I don't know what it will turn into."

When I was seven years old I had the misfortune to be laid up with scarlet fever, and then his gift of drawing was a great solace to me. The solitary days—for I was the first victim in the family—were very long, and I looked forward with intense interest to one half-hour after dinner, when he would come up and draw scenes from the history of a remarkable bull terrier and his family that went to the seaside, in a most human and child-delighting manner. I have seldom suffered a greater disappointment than when, one evening, I fell asleep just before this fairy half-hour, and lost it out of my life.

In those days he often used to take the three eldest of us out for a walk on Sunday afternoons, sometimes to the Zoological Gardens, more often to the lanes and fields between St. John's Wood and Hampstead or West End. For then the flood of bricks and mortar ceased on the Finchley Road just beyond the Swiss Cottage, and the West End Lane, winding solitary between its high hedges and rural ditches, was quite like a country road in holiday time, and was sometimes gladdened in June with real dog-roses, although the church and a few houses had already begun to encroach on the open fields at the end of the Abbey Road.

My father often used to delight us with sea stories and tales of animals, and occasionally with geological sketches suggested by the gravels of Hampstead Heath. But regular "shop" he would not talk to us, contrary to the expectation of people who have often asked me whether we did not receive quite a scientific training from his companionship.

At the Christmas dinner he invariably delighted the children by carving wonderful beasts, generally pigs, out of orange peel. When the marriage of his eldest daughter had taken her away from this important function, she was sent the best specimen as a reminder.

4 MARLBOROUGH PLACE,
Dec. 25, 1878.

DEAREST JESS—We have just finished the mid-day Christmas dinner, at which function you were badly wanted. The inflammation of the pudding was highly successful—in fact Vesuvian not to say Ætnaic—and I have never yet attained so high a pitch in piggygenesis as on this occasion.

The specimen I enclose, wrapped in a golden cerecloth, and with the remains of his last dinner in the proper region, will prove to you the heights to which the creative power of the true artist may soar. I call it a " Piggurne, or a Harmony in Orange and White."

Preserve it, my dear child, as evidence of the paternal genius, when those light and fugitive productions which are buried in the philosophical transactions and elsewhere are forgotten.

My best wishes to Fred and you, and may you succeed better than I do in keeping warm.—Ever your loving father, T. H. HUXLEY.

Later on, however, the younger children who kept up the home at Marlborough Place after the elder ones had married or gone out into the world, enjoyed more opportunities of his ever-mellowing companionship. Strongly as he upheld the conventions when these represented some valid results of social experience, he was always ready to set aside his mere likes and dislikes on good cause shown ; to follow reason as against the mere prejudice of custom, even his own.

Severe he might be on occasion, but never harsh. His idea in bringing up his children was to accustom them as early as possible to a certain amount of independence, at the same time trying to make them regard him as their best friend.

This aspect of his character is specially touched upon by Sir Leslie Stephen, in a letter written to my mother in July 1895 :—

No one, I think, could have more cordially admired Huxley's intellectual vigour and unflinching honesty than I. It pleases me to remember that I lately said something of this to him, and that he received what I said most heartily and kindly. But what now dwells most in my mind is the memory of old kindness, and of the days when I used to see him with you and his children. I may safely say that I never came from your house without thinking how good he is ; what a tender and affectionate nature the man has ! It did me good simply to see him. The recollection is sweet to me now, and I rejoice to think how infinitely better you know what I must have been dull indeed not more or less to perceive.

As he wrote to his son on his twenty-first birthday :—

You will have a son some day yourself, I suppose, and if you do, I can wish you no greater satisfaction than to be able to say that he has reached manhood without having given you a serious anxiety, and that you can look forward with entire confidence to his playing the man in the battle of life. I have tried to make you feel your responsibilities and act independently as early as possible—but, once for all, remember that I am not only your father but your nearest friend, ready to help you in all things reasonable, and perhaps in a few unreasonable.

This domestic happiness which struck others so forcibly was one of the vital realities of his existence. Without it his quick spirit and nervous temperament could never have endured the long and often embittered struggle—not merely with equanimity, but with a constant growth of sympathy for earnest humanity, which, in early days obscured from view by the turmoil of strife, at length became apparent to all as the tide of battle subsided. None realised more than himself what the sustaining help and comradeship of married life had wrought for him, alike in making his life worth living and in making his life's work possible. Here he found the pivot of his happiness and his strength; here he recognised to the full the care that took upon itself all possible burdens and left his mind free for his greater work.

He had always a great tenderness for children. "One of my earliest recollections of him," writes Jeffery Parker, "is in connection with a letter he wrote to my father, on the occasion of the death, in infancy, of one of my brothers. 'Why,' he wrote, 'did you not tell us before that the child was

named after me, that we might have made his short
life happier by a toy or two.' I never saw a man
more crushed than he was during the dangerous
illness of one of his daughters, and he told me that,
having then to make an after-dinner speech, he
broke down for the first time in his life, and for
one painful moment forgot where he was and what
he had to say. I can truly say that I never knew a
man whose way of speaking of his family, or whose
manner in his own home, was fuller of a noble,
loving, and withal playful courtesy."

After he had retired to Eastbourne, his grand-
children reaped the benefit of his greater leisure.
In his age his love of children brimmed over with
undiminished force, unimpeded by circumstances.
He would make endless fun with them, until one
little mite, on her first visit, with whom her grandfather was trying to ingratiate himself with a vast
deal of nonsense, exclaimed: "Well, you are the
curious'test old man I ever seen."

Another, somewhat older, developed a great liking
for astronomy under her grandfather's tuition. One
day a visitor, entering unexpectedly, was astonished
to find the pair of them kneeling on the floor in the
hall before a large sheet of paper, on which the
professor was drawing a diagram of the solar system
on a large scale, with a little pellet and a large ball
to represent earth and sun, while the child was
listening with the closest attention to an account
of the planets and their movements, which he knew

so well how to make simple and precise without ever being dull.

Children seemed to have a natural confidence in the expression of mingled power and sympathy which, especially in his later years, irradiated his "square, wise, swarthy face,"[1] and proclaimed to all the sublimation of a broad native humanity tried by adversity and struggle in the pursuit of noble ends. It was the confidence that an appeal would not be rejected, whether for help in distress, or for the satisfaction of the child's natural desire for knowledge.

Spirit and determination in children always delighted him. His grandson Julian, a curly-haired rogue, alternately cherub and pickle, was a source of great amusement and interest to him. The boy must have been about four years old when my father one day came in from the garden, where he had been diligently watering his favourite plants with a big hose, and said : " I like that chap ! I like the way he looks you straight in the face and disobeys you. I told him not to go on the wet grass again. He just looked up boldly, straight at me, as much as to say, ' What do *you* mean by ordering me about ?' and deliberately walked on to the grass."

The disobedient youth who so charmed his grand-

---

[1] "There never was a face, I do believe" (wrote Sir Walter Besant of the portrait by John Collier), "wiser, more kindly, more beautiful for wisdom and the kindliness of it, than this of Huxley."—The *Queen*, Nov. 16, 1895.

father's heart was the prototype of Sandy in Mrs.
Humphry Ward's *David Grieve.* When the book
came out my father wrote to the author: "We are
very proud of Julian's apotheosis.   He is a most
delightful imp, and the way in which he used to
defy me on occasion, when he was here, was quite
refreshing.   The strength of his conviction that
people who interfere with his freedom are certainly
foolish, probably wicked, is quite Gladstonian."

A year after, when Julian had learned to write,
and was reading the immortal *Water Babies,* wherein
fun is poked at his grandfather's name among the
authorities on water-babies and water-beasts of every
description, he greatly desired more light as to the
reality of water - babies.   There is a picture by
Linley Sambourne, showing my father and Owen
examining a bottled water-baby under big magnifying
glasses.   Here, then, was a real authority to consult.
So he wrote a letter of inquiry, first anxiously asking
his mother if he would receive in reply a "proper
letter" that he could read for himself, or a "wrong
kind of letter" that must be read to him.

DEAR GRANDPATER—Have you seen a Waterbaby?
Did you put it in a bottle? Did it wonder if it could
get out?   Can I see it some day?—Your loving
JULIAN.

To this he received the following reply from his
grandfather, neatly printed, letter by letter, very
unlike the orderly confusion with which his pen
usually rushed across the paper—time being so short

for such a multitude of writing—to the great per-
plexity, often, of his foreign correspondents.

HODESLEA.
STAVELEY ROAD.
EASTBOURNE

March 24
1892.

My dear Julian

. I never could make
sure about that Water
Baby. I have seen
Babies in water and
Babies in bottles; but
the Baby in the water
was not in a bottle and

the Baby in the bottle was
not in water.

My friend who wrote the
story of the Water Baby,
was a very kind man
and very clever. Perhaps
he thought I could
see as much in the
water as he did—

There are some people

who see a great deal

and some who see very

little in the same things.

When you grow up

I dare say you will be

one of the great-deal seers

and see things more

wonderful than Water

Babies where other folks

can see nothing

Give my best love to Daddy + Mammy and Trevenen — Grandmoo is a little better but not up yet —

Ever your lovung

Grandpater

Others of his family would occasionally receive elaborate pieces of nonsense, of which I give a couple of specimens. The following is to his youngest daughter :—

ATHENÆUM CLUB,
May 17, 1892.

DEAREST BABS—As I was going along Upper Thames
primary parenthesis
Street just now, I saw between Nos. 170 and 211 (but

you would like to know what I was going along that odorous street for. Well, it was to inquire how the
2nd p.
pen with which I am now writing—(you see it is a new-fangled fountain pen, warranted to cure the worst writing
2nd p.
and always spell properly)—works, because it would not work properly this morning. And the nice young woman
3rd p.   3rd p.
who took it from me—(as who should say you old foodle !)
4th p.
inked her own fingers enormously (which I told her I
4th p.
was pleased they were her fingers rather than mine)—
5th p.
But she only smole. (Close by was another shop where
6 or 7 p.   n.p.
they sold hose—(indiarubber, not knitted)—(and warranted to let water through, not keep it out); and I asked for a garden syringe, thinking such things likely to be kept by hosiers of that sort—and they said they
n.n.p.
had not any, but found they had a remnant cheap (price 3s.) which is less than many people pay for the other
end of pp.
hosiers' hose) a doorpost at the side of the doorway of some place of business with this remarkable notice : RULING GIRLS WANTED.

Don't you think you had better apply at once ? Jack will give you a character, I am sure, on the side of the art of ruling, and I will speak for the science—also of hereditary (on mother's side) instinct.

Well I am not sure about the pen yet—but there is no room for any more.—Ever your loving   DAD.

Epistolary composition on the model of a Gladstonian speech to a deputation on women's suffrage.

The other is to his daughter, Mrs. Harold Roller, who had sent him from abroad a friend's autograph-book for a signature :—

HODESLEA, EASTBOURNE,
Nov. 1, 1893.

The epistle of Thomas to the woman of the house of Harold.

1. I said it was an autograph-book ; and so it was.
2. And naughty words came to the root of my tongue.
3. And the recording angel dipped his pen in the ink and squared his elbows to write.
4. But I spied the hand of the lovely and accomplished but vagabond daughter.
5. And I smole ; and spoke not ; nor uttered the naughty words.
6. So the recording angel was sold ;
7. And was about to suck his pen.
8. But I said Nay ! give it to me.
9. And I took the pen and wrote on the book of the Autographs letters pleasant to the eye and easy to read.
10. Such as my printers know not : nor the postman— nor the correspondent, who riseth in his wrath and curseth over my epistle ordinary.

This to his youngest daughter, which, in jesting form, conveys a good deal of sound sense, was the sequel to a discussion as to the advisability of a University education for her own and another boy :—

HODESLEA, EASTBOURNE,
May 9, 1892.

DEAREST BABS—Bickers and Son have abased themselves, and assure me that they have fetched the Dicty. away and are sending it here. I shall believe them when it arrives.

As a rule, I do not turn up when I announce my coming, but I believe I shall be with you about dinner-time on Friday next (13th).

In the meanwhile, my good daughter, meditate these things :

1. Parents not too rich wish to send exceptionally clever, energetic lad to university—before taking up father's profession of architect.

2. E.c.e l. will be well taught classics at school—not well taught in other things—will easily get a scholarship either at school or university.   So much in parents' pockets.

3. E.c.e.l. will get as much mathematics, mechanics, and other needful preliminaries to architecture, as he wants (and a good deal more if he likes) at Oxford.   Excellent physical school there.

4. Splendid Art museums at Oxford.

5. Prigs not peculiar to Oxford.

6. Don Cambridge would choke science (except mathematics) if it could as willingly as Don Oxford and more so.

7. Oxford always represents English opinion, in all its extremes, better than Cambridge.

8. Cambridge better for doctors, Oxford for architects, poets, painters, and all that sort of cattle.

9. *Lawrence will go to Oxford* and become a real scholar, which is a great thing and a noble.   He will combine the new and the old, and show how much better the world would have been if it had stuck to Hellenism.   You are dreaming of the schoolboy who does not follow up his work, or becomes a mere poll man.   Good enough for parsons, not for men.   *Lawrence will go to Oxford.*—Ever your aggrawatin'   PA.

Like the old Greek sage and statesman, my father might have declared that old age found him ever learning.   Not indeed with the fiery earnestness of his young days of stress and storm ; but with the steady advance of a practised worker who cannot be unoccupied.   History and philosophy, especially biblical criticism, composed his chief reading in these later years.

Fortune had ceased her buffets ; broken health was restored ; and from his resting-place among his books and his plants he watched keenly the struggle which had now passed into other hands, still ready to strike a blow if need be, or even, on rare occasions, to return to the fighting line, as when he became a leader in the movement for London University reform.

His days at Eastbourne, then, were full of occupation, if not the occupation of former days.　The day began as early ; he never relaxed from the rule of an eight o'clock breakfast.　Then a pipe and an hour and a half of letter-writing or working at an essay. Then a short expedition around the garden, to inspect the creepers, tend the saxifrages, or see how the more exposed shrubs could best be sheltered from the shrivelling winds.　The gravelled terrace immediately behind the house was called the Quarterdeck ; it was the place for a brisk patrolling in uncertain weather or in a north wind.　In the lower garden was a parallel walk protected from the south by a high double hedge of cypress and golden elder, designed for shelter from the summer sun and southerly winds.

Then would follow another spell of work till near one o'clock; the weather might tempt him out again before lunch ; but afterwards he was certain to be out for an hour or two from half-past two.　However hard it blew, and Eastbourne is seldom still, the tiled walk along the sea-wall always offered the

possibility of a constitutional. But the high expanse
of the Downs was his favourite walk. The air of
Beachy Head, 560 feet up, was an unfailing tonic.
In the summer he used to keep a look-out for the
little flowers of the short, close turf of the chalk
which could remind him of his Alpine favourites, in
particular the curious phyteuma; and later on, in
the folds of the hills where he had marked them, the
English gentians.

After his walk, a cup of tea was followed by more
reading or writing till seven; after dinner another
pipe, and then he would return to my mother in the
drawing-room, and settle down in his particular arm-
chair, with some tough volume of history or theology
to read, every now and again scoring a passage for
future reference, or jotting a brief note on the margin.
At ten he would migrate to the study for a final smoke
before going to bed.

Such was his routine, broken by occasional visits
to town on business, for he was still Dean of the
Royal College of Science and a trustee of the British
Museum. Old friends came occasionally to stay for
a few days, and tea-time would often bring one or
two of the small circle of friends whom he had made
in Eastbourne. These also he occasionally visited,
but he scarcely ever dined out. The talking was too
tiring.

The change to Eastbourne cut away a whole
series of interests, but it imported a new and very
strong one into my father's life. His garden was

not only a convenient ambulatory, but, with its growing flowers and trees, became a novel and intense pleasure, until he began "to think with Candide that 'Cultivons notre jardin' comprises the whole duty of man."

It was strange that this interest should have come suddenly at the end of his life. Though he had won the prize in Lindley's botanical class, he had never been a field botanist till he was attracted by the Swiss gentians. As has been said before, his love of nature had never run to collecting either plants or animals. Mere "spider-hunters and hay-naturalists," as a German friend called them, he was inclined to regard as the camp-followers of science. It was the engineering side of nature, the unity of plan of animal construction, worked out in infinitely varying detail, which engrossed him. Walking once with Hooker in the Rhone valley, where the grass was alive with red and green grasshoppers, he said, "I would give anything to be as interested in them as you are."

But this feeling, unknown to him before, broke out in his gentian work. He told Hooker, "I can't express the delight I have in them." It continued undiminished when once he settled in the new house and laid out a garden. His especial love was for the rockery of Alpines, many of which came from Sir J. Hooker.

Here, then, he threw himself into gardening with characteristic ardour. He described his position as a kind of mean between the science of the botanist and the empiricism of the working gardener. He had plenty to suggest, but his gardener, like so many of his tribe, had a rooted mistrust of any gardening lore culled from books. "Books? They'll say anything in them books."   And he shared, moreover, that common superstition, perhaps really based upon a question of labour, that watering of flowers, unnecessary in wet weather, is actively bad in dry. So my father's chief occupation in the garden was to march about with a long hose, watering, and watering especially his alpines in the upper garden and along the terraces lying below the house. The saxifrages and the creepers on the house were his favourite plants.   When he was not watering the one he would be nailing up the other, for the winds of Eastbourne are remarkably boisterous, and shrivel up what they do not blow down.   "I believe I shall take to gardening," he writes, a few months after entering the new house, "if I live long enough. I have got so far as to take a lively interest in the condition of my shrubs, which have been awfully treated by the long cold."

From this time his letters contain many references to his garden. He is astonished when his gardener asks leave to exhibit at the local show, but delighted with his pluck. Hooker jestingly sends him a plant "which will flourish on any dry, neglected bit of wall, so I think it will just suit you."

Great improvements have been going on (he writes in 1892), and the next time you come you shall walk in the "avenue" of four box-trees. Only five are to be had for love or money at present, but there are hopes of a sixth, and then the "avenue" will be full ten yards long ! *Figurez vous ça !*

It was of this he wrote on October 1 :—

Thank Heaven we are settled down again and I can vibrate between my beloved books and even more beloved saxifrages.

The additions to the house are great improvements every way, outside and in, and when the conservatory is finished we shall be quite palatial ; but, alas, of all my box-trees only one remains green, that is the " amari," or more properly " fusci " aliquid.

Sad things will happen, however. Although the local florists vowed that the box-trees would not stand the winds of Eastbourne, he was set on seeing if he could not get them to grow despite the gardeners, whom he had once or twice found false prophets. But this time they were right. Vain were watering and mulching and all the arts of the husbandman. The trees turned browner and browner every day, and the little avenue from terrace to terrace had to be ignominiously uprooted and removed.

A sad blow this, worse even than the following :—

A lovely clematis in full flower, which I had spent hours in nailing up, has just died suddenly. I am more inconsolable than Jonah !

He answers some gardening chaff of Sir Michael Foster's :—

Wait till I cut you out at the Horticultural. I have
not made up my mind what to compete in yet. Look
out when I do !

And when the latter offered to propose him for
that Society, he replied :—

Proud an' 'appy should I be to belong to the Horticultural if you will see to it. Could send specimens of
nailing up creepers if qualification is required.

After his long battlings for his early loves of
science and liberty of thought, his later love of the
tranquil garden seemed in harmony with the dignified
rest from struggle.   To those who thought of the
past and the present, there was something touching
in the sight of the old man whose unquenched fires
now lent a gentler glow to the peaceful retirement
he had at length won for himself.   His latter days
were fruitful and happy in their unflagging intellectual
interests, set off by the new delights of the *succidia
altera*, that second resource of hale old age for many
a century.

All through his last and prolonged illness, from
earliest spring until midsummer, he loved to hear how
the garden was getting on, and would ask after certain
flowers and plants.   When the bitter cold spring was
over and the warm weather came, he spent most of
the day outside, and even recovered so far as to be
able to walk once into the lower garden and visit his
favourite flowers.   These children of his old age
helped to cheer him to the last.

APPENDIX I

As for this unfinished work, suggestive outlines left for others to fill in, Professor Howes writes to me in October 1899 :—

Concerning the papers at S.K. which, as part of the contents of your father's book-shelves, were given by him to the College, and now are arranged, numbered, and registered in order for use, there is evidence that in 1858 he, with his needles and eyeglass, had dissected and carefully figured the so-called pronephros of the Frog's tadpole, in a manner which as to accuracy of detail anticipated later discovery. Again, in the early '80's, he had observed and recorded in a drawing the præ-pulmonary aortic arch of the Amphibian, at a period antedating the researches of Boas, which in connection with its discovery placed the whole subject of the morphology of the pulmonary artery of the vertebrata on its final basis, and brought harmony into our ideas concerning it.

Both these subjects lie at the root of modern advances in vertebrate morphology.

Concerning the skull, he was in the '80's back to it with a will. His line of attack was through the lampreys and hags and the higher cartilaginous fishes, and he was following up a revolutionary conception (already hinted at in his Hunterian Lectures in 1864, and later in a Royal Society paper on *Amphioxus* in 1875), that the trabeculæ cranii, judged by their relationships to the

427

nerves, may represent a pair of præ-oral visceral arches. In his unpublished notes there is evidence that he was bringing to the support of this conclusion the discovery of a supposed 4th branch to the trigeminal nerve—the relationships of this (which he proposed to term the "hyporhinal" or palato-nasal division) and the ophthalmic (to have been termed the "orbitonasal" [1]) to the trabecular arch and a supposed præ-mandibular visceral cleft, being regarded as repetitional of those of the maxillary and mandibular divisions to the mandibular cleft. So far as I am aware, von Kupffer is the only observer who has given this startling conclusion support, in his famous *Studien* (Hf. I. Kopf Acipenser, München, 1893), and from the nature of other recent work on the genesis of parts of the cranium hitherto thought to be wholly trabecular in origin, it might well be further upheld. As for the discovery of the nerve, I have been lately much interested to find that Mr. E. Phelps Allis, jun., an investigator who has done grand work in Cranial Morphology, has recently and independently arrived at a similar result. It was while working in my laboratory in July last that he mentioned the fact to me. Remembering that your father had published the aforementioned hints on the subject, and recalling conversations I had with him, it occurred to me to look into his unpublished MSS. (then being sorted), if perchance he had gone further. And, behold ! there is a lengthy attempt to write the matter up in full, in which, among other things, he was seeking to show that, on this basis, the mode of termination of the notochord in the Craniata, and in the Branchiostomidæ (in which the trabecular arch is undifferentiated), is readily explained. Mr. Allis's studies are now progressing, and I have arranged with him that if, in the end, his results

---

[1] A term already applied by him in 1875 to the corresponding nerve in the Batrachia. (*Ency. Brit.* 9th edition, vol. i., art. 'Amphibia.")

come sufficiently close to your father's, he shall give his work due recognition and publicity.[1]

Among his schemes of the early '80's, there was actually commenced a work on the principles of Mammalian Anatomy and an Elementary Treatise on the Vertebrata. The former exists in the shape of a number of drawings with very brief notes, the latter to a slight extent only in MS. In the former, intended for the medical student and as a means of familiarising him with the anatomical "tree" as distinct from its surgical "leaves," your father once again returned to the skull, and he leaves a scheme for a revised terminology of its nerve exits worthy his best and most clear-headed endeavours of the past.[2] And well do I remember how, in the '80's, both in the class-room and in conversation, he would emphasise the fact that the hypoglossus nerve roots of the mammal arise serially with the ventral roots of the spinal nerves, little thinking that the discovery by Froriep, in 1886, of their dorsal ganglionated counterparts, would establish the actual homology between the two, and by leading to the conclusion that though actual vertebræ do not contribute to the formation of the mammalian skull, its occipital region is of truncal origin, mark the most revolutionary advance in cranial morphology since his own of 1856.

---

[1] See "The Lateral Sensory Canals, the Eye-Muscles, and the Peripheral Distribution of certain of the Cranial Nerves of Mustelus lævis" by Edward Phelps Allis, junr., reprinted from *Quart. Jour. Micr. Sci.* vol. xlv. part 2, New Series.

[2] Concerning this he wrote to Professor Howes in 1890 when giving him permission to denote two papers which he was about to present to the Zoological Society, as the first which emanated from the Huxley Research Laboratory :—"Pray do as you think best about the nomenclature. I remember when I began to work at the skull it seemed a hopeless problem, and years elapsed before I got hold of the clue."

And six weeks later, he writes :—"You are always welcome to turn anything of mine to account, though I vow I do not just now recollect anything about the terms you mention. If you were to examine me in my own papers, I believe I should be plucked."

Much of the final zoological work of his life lay with the Bony Fishes, and he leaves unfinished (indeed only just commenced) a memoir embodying a new scheme of classification of these, which shows that he was intending to do for them what he did for Birds in the most active period of his career. It was my good fortune to have helped as a hodman in the study of these creatures, with a view to a Text-book we were to have written conjointly, and as I realise what he was intending to make out of the dry facts, I am filled with grief at the thought of what we must have lost. His classification was based on the labours of years, as testified by a vast accumulation of rough notes and sketches, and as a conspicuous feature of it there stands the embodiment under one head of all those fishes having the swim-bladder in connection with the auditory organ by means of a chain of ossicles—a revolutionary arrangement, which later, in the hands of the late Dr. Sagemähl, and by his introduction of the famous term—" Ostariophyseæ," has done more than all else of recent years to clear the Ichthyological air. Your father had anticipated this unpublished, and in a proposal to unite the Herrings and Pikes into a single group, the "Clupesoces," he had further given promise of a new system, based on the study of the structure of the fins, jaws, and reproductive organs of the Bony Fishes, the classifications of which are still largely chaotic, which would have been as revolutionary as it was rational. New terms both in taxonomy and anatomy were contemplated, and in part framed. His published terms "Elasmo-" and "Cysto-arian" are the adjective form of two—far-reaching and significant—which give an idea of what was to have come. Similarly, the spinose fin-rays were to have been termed "acanthonemes," the branching and multiarticulate "arthronemes," and those of the more elementary and "adipose fin" type "protonemes": and had he lived to complete the task, I question whether it would not have excelled his earlier achievements.

The Rabbit was to have been the subject of the first of the aforementioned books, and in the desire to get at the full meaning of problems which arose during its progress, he was led to digress into a general anatomical survey of the Rodentia, and in testimony to this there remain five or six books of rough notes bearing dates 1880 to 1884, and a series of finished pencil-drawings, which, as works of art and accurate delineations of fact, are among the most finished productions of his hand. In the same manner his contemplated work upon the Vertebrata led him during 1879-1880 to renewed investigation of the anatomy of some of the more aberrant orders. Especially as concerning the Marsupialia and Edentata was this the case, and to the end in view he secured living specimens of the Vulpine Phalanger, and purchased of the Zoological Society the Sloths and Ant-eaters which during that period died in their Gardens. These he carefully dissected, and he leaves among his papers a series of incomplete notes (fullest as concerning the Phalanger and Cape Ant-eater [*Orycteropus*] [1]), which were never finished up.

They prove that he intended the production of special monographs on the anatomy of these peculiar mammalian forms, as he did on members of other orders which he had less fully investigated, and on the more important groups of fishes alluded to in the earlier part of my letter ; and there seems no doubt, from the collocation of dates and study of the order of the events, that his memorable paper " On the Application of the Laws of Evolution to the arrangement of the Vertebrata, and more particularly of the Mammalia," published in the *Proc. Zool. Soc.* for 1880,—the most masterly among his scientific theses— was the direct outcome of this intention, the only ex-

---

[1] I was privileged to assist in the dissection of the latter animal, and well do I remember how, when by means of a blow-pipe he had inflated the bladder, intent on determining its limit of distensibility, the organ burst, with unpleasant results, which called forth the remark, " I think we'll leave it at that ! "

pression which he gave to the world of the interaction of a series of revolutionary ideas and conceptions (begotten of the labours of his closing years as a working zoologist) which were at the period assuming shape in his mind. They have done more than all else of their period to rationalise the application of our knowledge of the Vertebrata, and have now left their mark for all time on the history of progress, as embodied in our classificatory systems.

He was in 1882 extending his important observations upon the respiratory apparatus from birds to reptiles, with results which show him to have been keenly appreciative of the existence of fundamental points of similarity between the Avian and Chelonian types—a field which has been more recently independently opened up by Milani.

Nor must it be imagined that after the publication of his ideal work on the Crayfishes in 1880, he had forsaken the Invertebrata. On the contrary, during the late '70's, and on till 1882, he accumulated a considerable number of drawings (as usual with brief notes), on the Mollusca. Some are rough, others beautiful in every respect, and among the more conspicuous outcomes of the work are some detailed observations on the nervous system, and an attempt to formulate a new terminology of orientation of the Acephalous Molluscan body. The period embraces that of his research upon the *Spirula* of the *Challenger* expedition, since published ; and incidentally to this he also accumulated a series of valuable drawings, with explanatory notes, of Cephalopod anatomy, which, as accurate records of fact, are unsurpassed.

As you are aware, he was practically the founder of the Anthropological Institute. Here again, in the late '60's and early '70's, he was most clearly contemplating a far-reaching inquiry into the physical anthropology of all races of mankind. There remain in testimony to this some 400 to 500 photographs (which I have had carefully arranged in order and registered), most of them of the nude figure standing erect, with the arm extended against a scale. A desultory correspondence proves that in connection with these he was in treaty with British residents and agents all over the world, with the Admiralty and naval officers, and that all was being done with a fixed idea in view. He was clearly contemplating something exhaustive· and definite which he never fulfilled, and the method is now the more interesting from its being essentially the same as that recently and independently adopted by Mortillet.

Beyond this, your father's notes reveal numerous other indications of matters and phases of activity, of great interest in their bearings on the history and progress of contemporary investigation, but these are of a detailed and wholly technical order.

APPENDIX II

His administrative work as an officer of the Royal Society is described in the following note by Sir Joseph Hooker :—

Mr. Huxley was appointed Joint-Secretary of the Royal Society, November 30, 1871, in succession to Dr. Sharpey, Sir George Airy being President, and Professor (now Sir George) Stokes, Senior Secretary. He held the office till November 30, 1880. The duties of the office are manifold and heavy ; they include attendance at all the meetings of the Fellows, and of the councils, committees, and sub-committees of the Society, and especially the supervision of the printing and illustrating all papers on biological subjects that are published in the Society's Transactions and Proceedings : the latter often involving a protracted correspondence with the authors. To this must be added a share in the supervision of the staff of officers, of the library and correspondence, and the details of house-keeping.

The appointment was well-timed in the interest of the Society, for the experience he had obtained as an officer in the Surveying Expedition of Captain Stanley rendered his co-operation and advice of the greatest value in the efforts which the Society had recently commenced to induce the Government, through the Admiralty especially, to undertake the physical and biological exploration of the ocean. It was but a few months before his appointment that he had been placed upon a committee of the Society, through which H.M.S. *Porcupine* was employed for this purpose in the European seas, and negotiations had already been commenced with the Admiralty for a voyage of circumnavigation with the same objects, which eventuated in the *Challenger* Expedition.

In the first year of his appointment, the equipment of the *Challenger*, and selection of its officers, was entrusted to the Royal Society, and in the preparation of the instructions to the naturalists Mr. Huxley had a dominating responsibility. In the same year a correspondence commenced with the India Office on the subject of deep-sea dredging in the Indian Ocean (it came to nothing), and another with the Royal Geographical Society on that of a North Polar Expedition, which resulted in the Nares Expedition (1875). In 1873, another with the Admiralty on the advisability of appointing naturalists to accompany two of the expeditions about to be despatched for observing the transit of Venus across the sun's disk in Mauritius and Kerguelen, which resulted in three naturalists being appointed. Arduous as was the correspondence devolving on the Biological Secretary, through the instructing and instalment of these two expeditions, it was as nothing compared with the official, demi-official, and private, with the Government and individuals, that arose from the Government request that the Royal Society should arrange for the publication and distribution of the enormous collections brought home by the above-named expedition. It is not too much to say that Mr. Huxley had a voice in every detail of these publications. The sittings of the Committee of Publication of the *Challenger* Expedition collections (of which Sir J. D. Hooker was chairman, and Mr. Huxley the most active member) were protracted from 1876 to 1895, and resulted in the publication of fifty royal quarto volumes, with plates, maps, sections, etc., the work of seventy-six authors, every shilling of the expenditure on which (some £50,000) was passed under the authority of the Committee of Publication.

Nor was Mr. Huxley less actively interested in the domestic affairs of the Society. In 1873 the whole establishment was translated from the building subsequently occupied by the Royal Academy to that which it now inhabits in the same quadrangle; a flitting of library stuff and appurtenances involving great responsibilities on the officers for the satisfactory re-establishment of the whole institution. In 1874 a very important alteration of the bye-laws was effected, whereby that which gave to Peers the privilege of being proposed for election as Fellows, without previous selection by the Committee (and to which bye-laws, as may be supposed, Mr. Huxley was especially repugnant), was replaced by one restricting that privilege to Privy Councillors. In 1875 he actively supported a proposition for extending the interests taken in the Society by holding annually a reception, to which the lady friends of the Fellows who were interested in science should be invited to inspect an exhibition of some of the more recent inventions, appliances, and discoveries in science. And in the same year another reform took place in which he was no less interested, which was the abolition of the entrance fees for ordinary Fellows, which had proved a bar to the coming forward of men of small incomes, but great eminence. The loss of income to the Society from this was met by a subscription of no less than £10,666, raised almost entirely amongst the Fellows themselves for the purpose.

In 1876 a responsibility, that fell heavily on the Secretaries, was the allotment annually of a grant by the Treasury of £4000, to be expended, under the direction of the Royal [1] and other learned societies, on the advancement of science. Every detail of the business of this grant is undertaken by a large committee of the Royal and other scientific societies, which meets in the Society's rooms, and where all the business connected with the grant is conducted and the records kept.

APPENDIX III

LIST OF ESSAYS, BOOKS, AND SCIENTIFIC MEMOIRS, BY T. H. HUXLEY

ESSAYS

" The Darwinian Hypothesis." (*Times*, December 26, 1859.) *Collected Essays*, ii.

" On the Educational Value of the Natural History Sciences." (An Address delivered at St. Martin's Hall, on July 22, 1854, and published as a pamphlet in that year.) *Lay Sermons ; Collected Essays*, iii.

" Time and Life." (*Macmillan's Magazine*, December 1859.)

" The Origin of Species." (The *Westminster Review*, April 1860.) *Lay Sermons ; Collected Essays*, ii.

" A Lobster : or the Study of Zoology." (A Lecture delivered at the South Kensington Museum in 1861,

---

[1] It is often called a grant to the Royal Society. This is an error. The Royal Society, as such, in no way participates in this grant. The Society makes grants from funds in its own possession only.

and subsequently published by the Department of Science and Art. Original title, " On the Study of Zoology.") *Lay Sermons; Collected Essays*, viii.

" Geological Contemporaneity and Persistent Types of Life." (The Anniversary Address to the Geological Society for 1862.) *Lay Sermons; Collected Essays*, viii.

" Six Lectures to Working Men on Our Knowledge of the Causes of the Phenomena of Organic Nature, 1863." *Collected Essays*, ii.

" Man's Place in Nature," *see* List of Books. Republished, *Collected Essays*, vii.

" Criticisms on ' The Origin of Species.' " (The *Natural History Review*, 1864.) *Lay Sermons ; Collected Essays*, iii.

" Emancipation—Black and White." (The *Reader*, May 20, 1865.) *Lay Sermons; Collected Essays*, iii.

" On the Methods and Results of Ethnology." (The *Fortnightly Review*, 1865.) *Critiques and Addresses ; Collected Essays*, vii.

" On the Advisableness of Improving Natural Knowledge." (A Lay Sermon delivered in St. Martin's Hall, January 7, 1866, and subsequently published in the *Fortnightly Review.*) *Lay Sermons ; Collected Essays*, i.

" A Liberal Education : and where to find it." (An Address to the South London Working Men's College, delivered January 4, 1868, and subsequently published in *Macmillan's Magazine.*) *Lay Sermons; Collected Essays*, iii.

" On a Piece of Chalk." (A Lecture delivered to the working men of Norwich, during the meeting of the British Association, in 1868. Subsequently published in *Macmillan's Magazine.*) *Lay Sermons; Collected Essays*, viii.

" On the Physical Basis of Life." (A Lay Sermon, delivered in Edinburgh, on Sunday, November 8,

1868, at the request of the late Rev. James Cranbrook ; subsequently published in the *Fortnightly Review.*) *Lay Sermons ; Collected Essays,* i.

" The Scientific Aspects of Positivism." (A Reply to Mr. Congreve's Attack upon the Preceding Paper. Published in the *Fortnightly Review,* 1869.) *Lay Sermons.*

" The Genealogy of Animals." (A Review of Haeckel's *Natürliche Schöpfungs - Geschichte.* The *Academy,* 1869.) *Critiques and Addresses ; Collected Essays,* ii.

" Geological Reform." (The Anniversary Address to the Geological Society for 1869.) *Lay Sermons ; Collected Essays,* viii.

" Scientific Education : Notes of an After-Dinner Speech." (Delivered before the Liverpool Philomathic Society in April 1869, and subsequently published in *Macmillan's Magazine.*) *Lay Sermons ; Collected Essays,* iii.

" On Descartes' ' Discourse touching the Method of using one's Reason rightly, and of seeking Scientific Truth.' " (An Address to the Cambridge Young Men's Christian Society, delivered on March 24, 1870, and subsequently published in *Macmillan's Magazine.*) *Lay Sermons ; Collected Essays,* i.

" On some Fixed Points in British Ethnology." (The *Contemporary Review,* July 1870.) *Critiques and Addresses ; Collected Essays,* vii.

" Biogenesis and Abiogenesis." (The Presidential Address to the British Association for the Advancement of Science, 1870.) *Critiques and Addresses ; Collected Essays,* viii.

" Paleontology and the Doctrine of Evolution." (The Presidential Address to the Geological Society, 1870.) *Critiques and Addresses ; Collected Essays,* viii.

" On Medical Education." (An Address to the Students of the Faculty of Medicine in University College, London, 1870.) *Critiques and Addresses ; Collected Essays,* iii.

"On Coral and Coral Reefs." (*Good Words*, 1870.) *Critiques and Addresses.*

"The School Boards: What they can do, and what they may do." (The *Contemporary Review*, December 1870.) *Critiques and Addresses; Collected Essays*, iii.

"Administrative Nihilism." (An Address delivered to the Members of the Midland Institute, on October 9, 1871, and subsequently published in the *Fortnightly Review*.) *Critiques and Addresses; Collected Essays*, i.

"Mr. Darwin's Critics." (The *Contemporary Review*, November 1871.) *Critiques and Addresses; Collected Essays*, ii.

"On the Formation of Coal." (A Lecture delivered before the Members of the Bradford Philosophical Institution, December 29, 1871, and subsequently published in the *Contemporary Review*.) *Critiques and Addresses; Collected Essays*, viii.

"Yeast." (The *Contemporary Review*, December 1871.) *Critiques and Addresses; Collected Essays*, viii.

"Bishop Berkeley on the Metaphysics of Sensation." (*Macmillan's Magazine*, June 1871.) *Critiques and Addresses; Collected Essays*, vi.

"The Problems of the Deep Sea" (1873). *Collected Essays*, viii.

"Universities: Actual and Ideal." (The Inaugural Address of the Lord Rector of the University of Aberdeen, February 27, 1874. *Contemporary Review*, 1874.) *Science and Culture; Collected Essays*, iii.

"Joseph Priestley." (An Address delivered on the Occasion of the Presentation of a Statue of Priestley to the Town of Birmingham on August 1, 1874.) *Science and Culture; Collected Essays*, iii.

"On the Hypothesis that Animals are Automata, and its History." (An Address delivered at the Meeting of the British Association for the Advancement of Science, at Belfast, 1874.) *Science and Culture; Collected Essays*, i.

"On some of the Results of the Expedition of H.M.S. *Challenger*," 1875. *Collected Essays*, viii.

"On the Border Territory between the Animal and Vegetable Kingdoms." (An Evening Lecture at the Royal Institution, Friday, January 28, 1876. *Macmillan's Magazine*, 1876.) *Science and Culture; Collected Essays*, viii.

"Three Lectures on Evolution." (New York, September 18, 20, 22, 1876.) *American Addresses; Collected Essays*, iv.

"Address on University Education." (Delivered at the opening of the Johns Hopkins University, Baltimore, September 12, 1876.) *American Addresses; Collected Essays*, iii.

"On the Study of Biology." (A Lecture in connection with the Loan Collection of Scientific Apparatus at South Kensington Museum, December 16, 1876.) *American Addresses; Collected Essays*, iii.

"Elementary Instruction in Physiology." (Read at the Meeting of the Domestic Economy Congress at Birmingham, 1877.) *Science and Culture; Collected Essays*, iii.

"Technical Education." (An Address delivered to the Working Men's Club and Institute, December 1, 1877.) *Science and Culture; Collected Essays*, iii.

"Evolution in Biology." (The *Encyclopædia Britannica*, ninth edition, vol. viii. 1878.) *Science and Culture; Collected Essays*, ii.

"Hume," 1878. *Collected Essays*, vi. See also under "Books."

"On Sensation and the Unity of Structure of the Sensiferous Organs." (An Evening Lecture at the Royal Institution, Friday, March 7, 1879.) *Nineteenth Century*, April, 1879. *Science and Culture; Collected Essays*, vi.

"Prefatory Note to the Translation of E. Haeckel's Freedom in Science and Teaching," 1879. (Kegan Paul.)

"On Certain Errors respecting the Structure of the Heart attributed to Aristotle." *Nature*, November 6, 1879. *Science and Culture.*

" The Coming of Age of ' The Origin of Species.' " (An Evening Lecture at the Royal Institution, Friday, April 9, 1880.) *Science and Culture; Collected Essays*, ii.

" On the Method of Zadig." (A Lecture delivered at the Working Men's College, Great Ormond Street, 1880. *Nineteenth Century*, June 1880.) *Science and Culture; Collected Essays*, iv.

" Science and Culture." (An Address delivered at the Opening of Sir Josiah Mason's Science College, at Birmingham, on October 1, 1880.) *Science and Culture; Collected Essays*, iii.

" The Connection of the Biological Sciences with Medicine." (An Address delivered at the Meeting of the International Medical Congress in London, August 9, 1881.) *Science and Culture; Collected Essays*, iii.

" The Rise and Progress of Paleontology." (An Address delivered at the York Meeting of the British Association for the Advancement of Science, 1881.) *Controverted Questions; Collected Essays*, iv.

" Charles Darwin." (Obituary Notice in *Nature*, April 1882.) *Collected Essays*, ii.

" On Science and Art in Relation to Education." (An Address to the Members of the Liverpool Institution, 1882.) *Collected Essays*, iii.

" The State and the Medical Profession." (The Opening Address at the London Hospital Medical School, 1884.) *Collected Essays*, iii.

" The Darwin Memorial." (A Speech delivered at the Unveiling of the Darwin Statue at South Kensington, June 9, 1885.) *Collected Essays*, ii.

" The Interpreters of Genesis and the Interpreters of Nature." (*Nineteenth Century*, December 1885.) *Controverted Questions; Collected Essays*, iv.

" Mr. Gladstone and Genesis." (*Nineteenth Century*,

February 1886.)   *Controverted   Questions ;   Collected Essays*, iv.

" The Evolution of Theology : An Anthropological Study." (*Nineteenth Century*, March and April 1886.)   *Controverted Questions ; Collected Essays*, iv.

" Science and Morals." (*Fortnightly Review*, November 1886.)   *Controverted Questions ; Collected Essays*, ix.

" Scientific and Pseudo-Scientific Realism." (*Nineteenth Century*, February 1887.)   *Controverted Questions ; Collected Essays*, v.

" Science and Pseudo-Science." (*Nineteenth Century*, April 1887.)   *Controverted Questions; Collected Essays*, v.

" An Episcopal Trilogy." (*Nineteenth Century*, November 1887.)   *Controverted Questions ; Collected Essays*, v.

" Address on behalf of the National Association for the Promotion of Technical Education " (1887).   *Collected Essays*, iii.

" The Progress of Science " (1887).   (Reprinted from *The Reign of Queen Victoria*, by T. H. Ward.)   *Collected Essays*, i.

" Darwin Obituary." (*Proc. Roy. Soc.* 1888.)   *Collected Essays*, ii.

" The Struggle for Existence in Human Society." (*Nineteenth Century*, February 1888.)   *Collected Essays*, ix.

" Agnosticism." (*Nineteenth Century*, February 1889.)   *Controverted Questions ; Collected Essays*, v.

" The Value of Witness to the Miraculous." (*Nineteenth Century*, March 1889.)   *Controverted Questions ; Collected Essays*, v.

" Agnosticism : A Rejoinder." (*Nineteenth Century*, April 1889.)   *Controverted Questions ; Collected Essays*, v.

" Agnosticism and Christianity." (*Nineteenth Century*, June 1889.)   *Controverted Questions; Collected Essays*, v.

" The Natural Inequality of Men." (*Nineteenth Century*, January 1890.)   *Collected Essays*, i.

" Natural Rights and Political Rights." (*Nineteenth Century*, February 1890.)   *Collected Essays*, i.

"Capital, the Mother of Labour." (*Nineteenth Century,* March 1890.) *Collected Essays,* ix.

"Government : Anarchy or Regimentation." (*Nineteenth Century,* May 1890.) *Collected Essays,* i.

"The Lights of the Church and the Light of Science." (*Nineteenth Century,* July 1890.) *Controverted Questions ; Collected Essays,* iv.

"The Aryan Question." (*Nineteenth Century,* November 1890.) *Collected Essays,* vii.

"The Keepers of the Herd of Swine." (*Nineteenth Century,* December 1890.) *Controverted Questions ; Collected Essays,* v.

"Autobiography." (1890, *Collected Essays,* i.) This originally appeared with a portrait in a series of biographical sketches by C. Engel.

"Illustrations of Mr. Gladstone's Controversial Methods." (*Nineteenth Century,* March 1891). *Controverted Questions ; Collected Essays,* v.

"Hasisadra's Adventure." (*Nineteenth Century,* June 1891.) *Controverted Questions ; Collected Essays,* iv.

"Possibilities and Impossibilities." (The *Agnostic Annual* for 1892.) 1891, *Collected Essays,* v.

"Social Diseases and Worse Remedies." (1891.) Letters to the *Times,* December 1890 and January 1891. Published in pamphlet form (Macmillan & Co.) 1891. *Collected Essays,* ix.

"An Apologetic Irenicon." (*Fortnightly Review,* November 1892.)

"Prologue to 'Controverted Questions'" (1892). *Controverted Questions ; Collected Essays,* v.

"Evolution and Ethics," being the Romanes Lecture for 1893. Also "Prolegomena," 1894. *Collected Essays,* ix.

"Owen's Position in the History of Anatomical Science," being a chapter in the *Life of Sir Richard Owen,* by his grandson, the Rev. Richard Owen (1894). *Sci. Mem.* iv.

BOOKS

" Kölliker's Manual of Human Histology. (Translated and edited by T. H. Huxley and G. Busk), 1853.

" Evidence as to Man's Place in Nature," 1863.

" Lectures on the Elements of Comparative Anatomy " (one volume only published), 1864.

" Elementary Atlas of Comparative Osteology " (in 12 plates), 1864.

" Lessons in Elementary Physiology." First edition printed 1866 ; second edition, 1868 ; reprinted 1869, 1870, 1871, 1872 (twice) ; third edition, 1872 ; reprinted 1873, 1874, 1875, 1876, 1878, 1879, 1881, 1883, 1884 (six times) ; fourth edition, 1885 ; reprinted 1886, 1888, 1890, 1892, 1893 (twice), 1896, 1898.

" An Introduction to the Classification of Animals," 1869.

" Lay Sermons, Addresses, and Reviews." First edition printed 1870 ; second edition, 1871 ; reprinted 1871, 1872, 1874, 1877, 1880, 1883 ; third edition, 1887 ; reprinted 1891, 1893 (twice), 1895, 1899.

" Essays Selected from Lay Sermons, Addresses, and Reviews." First edition, 1871 ; reprinted 1874, 1877.

" Manual of the Anatomy of Vertebrated Animals," 1871 (Churchill).

" Critiques and Addresses." First edition printed 1873 ; reprinted 1883 and 1890.

" A Course of Practical Instruction in Elementary Biology." By Prof. Huxley and Dr. H. N. Martin. First edition printed 1875 ; second edition, 1876 ; reprinted 1877 (twice), 1879 (twice), 1881, 1882, 1883, 1885, 1886 (three times), 1887 ; third edition, edited by Messrs. Howes and Scott, 1887 ; reprinted 1889, 1892, 1898.

"American Addresses." First edition printed 1877 ; reprinted 1886.

"Anatomy of Invertebrated Animals," 1877.

"Physiography." First edition, 1877 ; reprinted 1877, 1878, 1879, 1880, 1881, 1882, 1883, 1884, 1885 (three times), 1887, 1888, 1890, 1891, 1893, 1897.

"Hume." English Men of Letters Series. First edition printed 1878 ; reprinted 1879 (twice), 1881, 1886, 1887, 1895.

"The Crayfish : an Introduction to the Study of Zoology," 1879.

"Evolution and Ethics." First edition printed 1893 ; reprinted 1893 (three times); second edition, 1893 ; third edition, 1893 ; reprinted 1894.

"Introductory Science Primer." First edition printed 1880 ; reprinted 1880, 1886, 1888, 1889 (twice), 1893, 1895, 1899.

"Science and Culture, and other Essays." First edition printed 1881 ; reprinted 1882, 1888.

"Social Diseases and Worse Remedies." First edition printed 1891 ; reprinted, with additions, 1891 (twice).

"Essays on some Controverted Questions." Printed in 1892.

Collected Essays. Vol. I. "Method and Results." First edition printed 1893 ; reprinted 1894, 1898.

Vol. II. "Darwiniana." First edition printed 1893 ; reprinted 1894.

Vol. III. "Science and Education." First edition printed 1893 ; reprinted 1895.

Vol. IV. "Science and Hebrew Tradition." First edition printed 1893 ; reprinted 1895, 1898.

Vol. V. "Science and Christian Tradition." First edition printed 1894 ; reprinted 1895, 1897.

Vol. VI. "Hume, with Helps to the Study of Berkeley." First edition printed 1894 ; reprinted 1897.

Vol. VII. "Man's Place in Nature." First printed for Macmillan and Co. in 1894 ; reprinted 1895, 1897.

Vol. VIII. "Discourses, Biological and Geological." First edition printed 1894 ; reprinted 1896.

Vol. IX. "Evolution and Ethics and other Essays." First edition printed 1894 ; reprinted 1895, 1898.

"Scientific Memoirs," vol. i. printed 1898, vol. ii. printed 1899, vol. iii. 1901, vol. iv. 1902.

SCIENTIFIC MEMOIRS

"On a Hitherto Undescribed Structure in the Human Hair Sheath," *Lond. Medical Gazette*, i. 1340 (July 1845).

"Examination of the Corpuscles of the Blood of Amphioxus Lanceolatus," *Brit. Assoc. Report* (1847), pt. ii. 95 ; *Sci. Mem.* i.

"Description of the Animal of Trigonia," *Proc. Zool. Soc.* vol. xvii. (1849), 30-32 ; also in *Ann. and Mag. of Nat. Hist.* v. (1850), 141-143 ; *Sci. Mem.* i.

"On the Anatomy and the Affinities of the Family of the Medusæ," *Phil. Trans. Roy. Soc.* (1849), pt. ii. 413 ; *Sci. Mem.* i.

"Notes on Medusæ and Polypes," *Ann. and Mag. of Nat. Hist.* vi. (1850), 66, 67 ; *Sci. Mem.* i.

"Observations sur la Circulation du Sang chez les Mollusques des Genres Firole et Atlante." (Extraites d'une lettre adressée à M. Milne-Edwards.) *Annales des Sciences Naturelles*, xiv. (1850), 193-195 ; *Sci. Mem.* i.

"Observations upon the Anatomy and Physiology of Salpa and Pyrosoma," *Phil. Trans. Roy. Soc.* (1851), pt. ii. 567-594 ; also in *Ann. and Mag. of Nat. Hist.* ix. (1852), 242-244 ; *Sci. Mem.* i.

"Remarks upon Appendicularia and Doliolum, two Genera of the Tunicata," *Phil. Trans. Roy. Soc.* (1851), pt. ii. 595-606 ; *Sci. Mem.* i.

" Zoological Notes and Observations made on board H.M.S.
   Rattlesnake during the years 1846-1850," *Ann. and
   Mag. Nat. Hist.* vii. ser. ii. (1851), 304-306, 370-374 ;
   vol. viii. 433-442 : *Sci. Mem.* i.
" Observations on the Genus Sagitta," *Brit. Assoc. Report*
   (1851), pt. ii. 77, 78 (sectional transactions) ; *Sci.
   Mem.* i.
" An Account of Researches into the Anatomy of the
   Hydrostatic Acalephæ," *Brit. Assoc. Report* (July 1851),
   pt. ii. 78-80 (sectional transactions) ; *Sci. Mem.* i.
" Description of a New Form of Sponge-like Animal,"
   Brit. Assoc. Report (July 1851), pt. ii. 80 (sectional
   transactions) ; *Sci. Mem.* i.
" Report upon the Researches of Prof. Müller into the
   Anatomy and Development of the Echinoderms,"
   Ann. and Mag. of Nat. Hist. ser. ii., vol. viii. (1851),
   1-19 ; *Sci. Mem.* i.
" Ueber die Sexualorgane der Diphydae und Physo-
   phoridae," Müller's *Archiv für Anatomie, Physiologie,
   und Wissenschaftliche Medicin* (1851), 380-384. *Sci.
   Mem.* i.
" Lacinularia Socialis : A Contribution to the Anatomy
   and Physiology of the Rotifera," *Trans. Micr. Soc.*,
   Lond., new series, i. (1853), 1-19 ; (Read December
   31, 1851). *Sci. Mem.* i.
" Upon Animal Individuality," *Proc. Roy. Inst.* i. (1851-
   54), 184-189. (Abstract of a Friday evening dis-
   course delivered on 30th April 1852.) *Sci.
   Mem.* i.
" On the Morphology of the Cephalous Mollusca, as Illus-
   trated by the Anatomy of certain Heteropoda and
   Pteropoda collected during the voyage of H.M.S.
   Rattlesnake in 1846-50," *Phil. Trans. Roy. Sci.* cxliii.
   (1853), part i. 29-66. *Sci. Mem.* i.
" Researches into the Structure of the Ascidians," *Brit.
   Assoc. Report* (1852), part ii. 76-77. *Sci. Mem.* i.
" On the Anatomy and Development of Echinococcus

Veterinorum," *Proc. Zool. Soc.* xx. (1852), 110-126. *Sci. Mem.* i.

" On the Identity of Structure of Plants and Animals " ; Abstract of a Friday evening discourse delivered at the Royal Institution on April 15, 1853 ; *Proc. Roy. Inst.* i. (1851-54), 298-302 ; *Edinburgh New Phil. Jour.* liii. (1852), 172-177. *Sci. Mem.* i.

" Observations on the Existence of Cellulose in the Tunic of Ascidians," *Quart. Jour. Micr. Sci.* i. 1853 ; *Sci. Mem.* i.

" On the Development of the Teeth, and on the Nature and Import of Nasmyth's ' Persistent Capsule,' " *Quart. Jour. Micr. Sci.* i. 1853. *Sci. Mem.* i.

" The Cell-Theory (Review)," *Brit. and For. Med. Chir. Review,* xii. (1853), 285-314. *Sci. Mem.* i.

" On the Vascular System of the Lower Annulosa," *Brit. Assoc. Report* (1854), part ii. p. 109. *Sci. Mem.* i.

" On the Common Plan of Animal Forms." (Abstract of a Friday evening discourse delivered at the Royal Institution on May 12, 1854.) *Proc. Roy. Inst.* i. (1851-54), 444-446. *Sci. Mem.* i.

" On the Structure and Relation of the Corpuscula Tactus (Tactile Corpuscles or Axile Corpuscles) and of the Pacinian Bodies," *Quart. Jour. Micr. Sci.* ii. (1853), 1-7. *Sci. Mem.* i.

" On the Ultimate Structure and Relations of the Malpighian Bodies of the Spleen and of the Tonsillar Follicles," *Quart. Jour. Micr. Sci.* ii. (1854), 74-82. *Sci. Mem.* i.

" On certain Zoological Arguments commonly adduced in favour of the Hypothesis of the Progressive Development of Animal Life in Time." (Abstract of a Friday evening discourse delivered on April 20, 1855.) *Proc. Roy. Inst.* ii. (1854 - 58), 82 - 85. *Sci. Mem.* i.

" On Natural History as Knowledge, Discipline, and Power," *Roy. Inst. Proc.* ii. (1854-58), 187-195.

(Abstract of a discourse delivered on Friday, February 15, 1856.) *Sci. Mem.* i.

"On the Present State of Knowledge as to the Structure and Functions of Nerve," *Proc. Roy. Inst.* ii. (1854-58), 432-437. (Abstract of a discourse delivered on Friday, May 15, 1857.) *Sci. Mem.* i.

(Translation) "On Tape and Cystic Worms," von Siebold (1857), for the Sydenham Society.

"Contributions to *Icones Zootomicæ*," by Victor Carus (1857).

"On the Phenomena of Gemmation." (Abstract of a discourse delivered on Friday, May 21, 1858.) *Proc. Roy. Inst.* ii. (1854-58), 534-538 ; *Silliman's Journal,* xxviii. (1859), 206-209. *Sci. Mem.* i.

"Contributions to the Anatomy of the Brachiopoda," *Proc. Roy. Soc.* vii. (1854-55), 106-117 ; 241, 242. *Sci. Mem.* i.

"On Hermaphrodite and Fissiparous Species of Tubicolar Annelidæ (Protula Dysteri)," *Edin. New Phil. Jour.* i. (1855), 113-129. *Sci. Mem.* i.

"On the Structure of Noctiluca Miliaris," *Quart. Jour. Micr. Soc.* iii. (1855), 49-54. *Sci. Mem.* i.

"On the Enamel and Dentine of the Teeth," *Quart. Jour. Micr. Soc.* iii. (1855), 127-130. *Sci. Mem.* i.

"Memoir on Physalia," *Proc. Linn. Soc.* ii. (1855), 3-5. *Sci. Mem.* i.

"On the Anatomy of Diphyes, and on the Unity of Composition of the Diphyidæ and Physophoridæ, etc.," *Proc. Linn. Soc.* ii. (1855), 67-69. *Sci. Mem.* i.

"Tegumentary Organs," *The Cyclopædia of Anatomy and Physiology,* edited by Robert B. Todd, M.D., F.R.S. (The fascicules containing this article were published between August 1855 and October 1856.) *Sci. Mem.* i.

"On the Method of Palæontology," *Ann. and Mag. of Nat. Hist.* xviii. (1856), 43-54. *Sci. Mem.* i.

"On the Crustacean Stomach," *Jour. Linn. Soc.* iv. 1856. (Never finally written.)

"Observations on the Structure and Affinities of Himan-topterus," *Quart. Jour. Geol. Soc.* xii. (1856), 34-37. *Sci. Mem.* i.

"Further Observations on the Structure of Appendicula Flabellum (Chamisso)," *Quart. Jour. Micr. Soc.* iv. (1856), 181-191. *Sci. Mem.* i.

"Note on the Reproductive Organs of the Cheilostome Polyzoa," *Quart. Jour. Micr. Soc.* iv. (1856), 191, 192. *Sci. Mem.* i.

"Description of a New Crustacean (Pygocephalus Cooperi, Huxley) from the Coal-measures," *Quart. Jour. Geol. Soc.* xiii. (1857), 363-369. *Sci. Mem.* i.

"On Dysteria, a New Genus of Infusoria," *Quart. Jour. Micr. Soc.* v. (1857), 78-82. *Sci. Mem.* i.

"Review of Dr. Hannover's Memoir : *Ueber die Entwicke-lung und den Bau des Säugethierzahns,*" *Quart. Jour. Micr. Soc.* v. (1857), 166-171. *Sci. Mem.* i.

"Letter to Mr. Tyndall on the Structure of Glacier Ice," *Phil. Mag.* xiv. (1857), 241-260. *Sci. Mem.* i.

"On Cephalaspis and Pteraspis," *Quart. Jour. Geol. Soc.* xiv. (1858), 267-280. *Sci. Mem.* i.

"Observations on the Genus Pteraspis," *Brit. Assoc. Report* (1858), part ii. 82, 83. *Sci. Mem.* i.

"On a New Species of Plesiosaurus (P. Etheridgii) from Street, near Glastonbury ; with Remarks on the Structure of the Atlas and the Axis Vertebræ and of the Cranium in that Genus," *Quart. Jour. Geol. Soc.* xiv. (1853), 281-94. *Sci. Mem.* i.

"On the Theory of the Vertebrate Skull," *Proc. Roy. Soc.* ix. (1857-59), 381-457 ; *Ann. and Mag. of Nat. Hist.* iii. (1859), 414-39. *Sci. Mem.* i.

"On the Structure and Motion of Glaciers," *Phil. Trans. Roy. Soc.* cxlvii. (1857), 327-346. (Received and read January 15, 1857.) *Sci. Mem.* ii.

"On the Agamic Reproduction and Morphology of Aphis," *Trans. Linn. Soc.* xxii. (1858), 193-220, 221-236. (Read November 5, 1857.) *Sci. Mem.* ii

"On Some Points in the Anatomy of Nautilus Pompilius,"
Jour. Linn. Soc. iii. (1859) (*Zool.*), 36-44. (Read
June 3, 1858.) *Sci. Mem.* ii.

"On the Persistent Types of Animal Life," *Proc. Roy.
Inst. of Great Britain,* iii. (1858-62), 151 153.
(Friday, June 3, 1859.) *Sci. Mem.* ii.

"On the Stagonolepis Robertsoni (Agassiz) of the Elgin
Sandstones; and on the Recently Discovered Foot-
marks in the Sandstones of Cummingstone," *Quart.
Jour. Geol. Soc.* xv. (1859), 440-460. *Sci. Mem.* ii.

"On Some Amphibian and Reptilian Remains from
South Africa and Australia," *Quart. Jour. Geol. Soc.*
xv. (1859), 642-649. (Read March 3, 1859.) *Sci.
Mem.* ii.

"On a New Species of Dicynodon (D. Murrayi) from
near Colesberg, South Africa; and on the Structure
of the Skull in the Dicynodonts," *Quart. Jour. Geol.
Soc.* xv. (1859), 649-658. (Read March 23, 1859.)
Sci. Mem. ii.

"On Rhamphorhynchus Bucklandi, a Pterosaurian from
the Stonesfield Slate," *Quart. Jour. Geol. Soc.* xv.
(1859), 658-670. (Read March 23, 1859.) *Sci.
Mem.* ii.

"On a Fossil Bird and a Fossil Cetacean from New
Zealand," *Quart. Jour. Geol. Soc.* xv. (1859), 670-677.
(Read March 23, 1859.) *Sci. Mem.* ii.

"On the Dermal Armour of Crocodilus Hastingsiæ,"
Quart. Jour. Geol. Soc. xv. (1859), 678-680. (Read
March 23, 1859.) *Sci. Mem.* ii.

"British Fossils," part i. "On the Anatomy and Affinities
of the Genus Pterygotus," *Mem. Geol. Sur. of United
Kingdom,* Monograph I. (1859), 1-36. *Sci. Mem.* ii.

"British Fossils," part ii. "Description of the Species of
Pterygotus," by J. W. Salter, F.G.S., A.L.S., *Mem.
Geol. Sur. of United Kingdom,* Monograph I. (1859),
37-105. *Sci. Mem.* ii.

"On Dasyceps Bucklandi (Labyrinthodon Bucklandi,

Lloyd)," *Mem. Geol. Sur. of United Kingdom* (1859), 52-56. *Sci. Mem.* ii.

" On a Fragment of a Lower Jaw of a Large Labyrinthodont from Cubbington," *Mem. Geol. Sur. of United Kingdom* (1859), 56-57. *Sci. Mem.* ii.

" Observations on the Development of Some Parts of the Skeleton of Fishes," *Quart. Jour. Micr. Sci.* vii. (1859), 33-46. *Sci. Mem.* ii.

" On the Dermal Armour of Jacare and Caiman, with Notes on the Specific and Generic Characters of Recent Crocodilia," *Jour. Linn. Soc.* iv. (1860) (*Zool.*), 1-28. (Read February 15, 1859.) *Sci. Mem.* ii.

" On the Anatomy and Development of Pyrosoma," *Trans. Linn. Soc.* xxiii. (1862), 193-250. (Read December 1, 1859.) *Sci. Mem.* ii.

" On the Oceanic Hydrozoa," *Ray Soc.* (1859).

" On Species and Races, and Their Origin " (1860), *Proc. Roy. Inst.* iii. (1858-62), 195-200 ; *Ann. and Mag. of Nat. Hist.* v. (1860), 344-346. *Sci. Mem.* ii.

" On the Structure of the Mouth and Pharynx of the Scorpion," *Quart. Jour. Micr. Sci.* viii. (1860), 250-254. *Sci. Mem.* ii.

" On the Nature of the Earliest Stages of the Development of Animals," *Proc. Roy. Inst.* iii. (1858-62), 315-317. (February 8, 1861.) *Sci. Mem.* ii.

" On a New Species of Macrauchenia (M. Boliviensis)," *Quart. Jour. Geol. Soc.* xvii. (1861), 73-84. *Sci. Mem.* ii.

" On Pteraspis Dunensis (Archæoteuthis Dunensis, Römer)," *Quart. Jour. Geol. Soc.* xvii. (1861), 163-166. *Sci. Mem.* ii.

" Preliminary Essay upon the Systematic Arrangement of the Fishes of the Devonian Epoch," *Mem. Geol. Sur. of United Kingdom*, " Figures and Descriptions of British Organic Remains " (1861, Decade *x*), 41-46. *Sci. Mem.* ii.

" Glyptolæmus Kinnairdi," *Mem. Geol. Sur. of United*

Kingdom, "Figures and Descriptions of British and Organic Remains" (1861, Decade x), 41-56. *Sci. Mem.* ii.

"Phaneropleuron Andersoni," *Mem. Geol. Sur. of United Kingdom*, "Figures and Descriptions of British Organic Remains" (1861, Dec. x), 47-49. *Sci. Mem.* ii.

"On the Zoological Relations of Man with the Lower Animals," *Nat. Hist. Rev.* (1861), 67-84. *Sci. Mem.* ii.

"On the Brain of Ateles Paniscus," *Proc. Zool. Soc.* (1861), 247-260. *Sci. Mem.* ii.

"On Fossil Remains of Man," *Proc. Roy. Inst.* (1858-62), 420-422. (February 7, 1862.) *Sci. Mem.* ii.

"Anniversary Address to the Geological Society, 1862," *Quart. Jour. Geol. Soc.* xviii. (1862), xl-liv. *See* also in list of Essays, "Geological Contemporaneity, etc." *Sci. Mem.* ii.

"On the New Labyrinthodonts from the Edinburgh Coalfield," *Quart. Jour. Geol. Soc.* xviii. (1862), 291-296. *Sci. Mem.* ii.

"On a Stalk-eyed Crustacean from the Carboniferous Strata near Paisley," *Quart. Jour. Geol. Soc.* xviii. (1862), 420-422. *Sci. Mem.* ii.

"On the Premolar Teeth of Diprotodon, and on a New Species of that Genus (D. Australis)," *Quart. Jour. Geol. Soc.* xviii. (1862), 422-427. *Sci. Mem.* ii.

"Description of a New Specimen of Glyptodon recently acquired by the Royal College of Surgeons," *Proc. Roy. Soc.* xii. (1862-63), 316-326. *Sci. Mem.* ii.

"Letter on the Human Remains found in Shell-mounds" (June 28, 1862), *Trans. Ethn. Soc.* ii. (1863), 265-266. *Sci. Mem.* ii.

"Description of Anthracosaurus Russelli, a New Labyrinthodont from the Lanarkshire Coal-field," *Quart. Jour. Geol. Soc.* xix. (1863), 56-68. *Sci. Mem.* ii.

"On the Form of the Placenta in the Cape Hyrax," *Proc. Zool. Soc.* (1863), p. 237. (The paper was never written in full; the materials and an unfinished drawing of the membranes are at South Kensington.)

" Further Remarks upon the Human Remains from the Neanderthal," *Nat. Hist. Rev.* (1864), 429-446. *Sci. Mem.* ii.

" On the Angwántibo (Arctocebus Calabarensis, Gray) of Old Calabar," *Proc. Zool. Soc.* (1864), 314-335. *Sci. Mem.* ii.

" On the Structure of the Skull of Man, the Gorilla, the Chimpanzee, and the Orang-Utan, during the period of the first dentition," *Proc. Zool. Soc.* (1864), p. 586. (This paper was never written in full, but was incorporated in " Man's Place in Nature.")

" On the Cetacean Fossils termed ' Ziphius ' by Cuvier, with a Notice of a New Species (Belemnoziphius Compressus) from the Red Crag," *Quart. Jour. Geol. Soc.* xx. (1864), 388-396. *Sci. Mem.* iii.

" On the Structure of the Belemnitidæ," *Mem. Geol. Sur. of United Kingdom,* Monograph II. (1864). *Sci. Mem.* iii.

" On the Osteology of the Genus Glyptodon " (1864), *Phil. Trans. Roy. Soc.* clv. (1865), 31-70. *Sci. Mem.* iii.

" On the Structure of the Stomach in Desmodus Rufus," *Proc. Zool. Soc.* (1865), 386-390. *Sci. Mem.* iii.

" On a Collection of Vertebrate Fossils from the Panchet Rocks, Ranigunj, Bengal," *Mem. Geol. Sur. of India ; Palæontologica Indica,* ser. iv. ; *Indian Pretertiary Vertebrata,* i. (1865-85). *Sci. Mem.* iii.

" On the Methods and Results of Ethnology " (1865), *Proc. Roy. Inst.* iv. (1866), 460-463. *Sci. Mem.* iii. See also *Collected Essays,* vii.

" Explanatory Preface to the Catalogue of the Palæontological Collection in the Museum of Practical Geology " (1865). *Sci. Mem.* iii. See *Principles and Methods of Paleontology,* 1869.

" On Two Extreme Forms of Human Crania," *Anthropological Review,* iv. (1866), 404-406.

" On a Collection of Vertebrate Remains from the Jarrow Colliery, Kilkenny, Ireland," *Geol. Mag.* iii. (1866), 165-171. *Sci. Mem.* iii.

"On some Remains of Large Dinosaurian Reptiles from the Stormberg Mountains, South Africa," *Phil. Mag.* xxxii. (1866), 474-475 ; *Quart. Jour. Geol. Soc.* xxiii. (1867), 1-6. *Sci. Mem.* iii.

"On a New Specimen of Telerpeton Elginense" (1866), *Quart. Jour. Geol. Soc.* xxiii. (1867), 77-84. *Sci. Mem.* iii.

"Notes on the Human Remains of Caithness" (1866), in the *Prehistoric Remains of Caithness*, by S. Laing.

"On Two Widely Contrasted Forms of the Human Cranium," *Jour. Anat. and Phys.* i. (1867), 60-77. *Sci. Mem.* iii.

"On Acanthopholis Horridus, a New Reptile from the Chalk-Marl," *Geol. Mag.* iv. (1867), 65-67. *Sci. Mem.* iii.

"On the Classification of Birds ; and on the Taxonomic Value of the Modifications of certain of the Cranial Bones observable in that Class," *Proc. Zool. Soc.* (1867), 415-472. *Sci. Mem.* iii.

"On the Animals which are most nearly Intermediate between Birds and Reptiles," *Ann. and Mag. of Nat. Hist.* ii. (1868), 66-75. *Sci. Mem.* iii.

"On Saurosternon Bainii and Pristerodon M'Kayi, two New Fossil Lacertilian Reptiles from South Africa," *Geol. Mag.* v. (1868), 201-205. *Sci. Mem.* iii.

"Reply to Objections on my Classification of Birds," *Ibis,* iv. (1868), 357-362.

"On the Form of the Cranium among the Patagonians and Fuegians, with some Remarks upon American Crania in general," *Jour. Anat. and Phys.* ii. (1868), 253-271. *Sci. Mem.* iii.

"On some Organisms living at Great Depths in the North Atlantic Ocean," *Quart. Jour. Micr. Sci.* viii. (1868), 203-212. *Sci. Mem.* iii.

"Remarks upon Archæopteryx Lithographica," *Proc. Roy. Soc.* xvi. (1868), 243-248. *Sci. Mem.* iii.

"On the Classification and Distribution of the Alectoromorphæ and Heteromorphæ," *Proc. Zool. Soc.* (1868), 294-319. *Sci. Mem.* iii.

"On Hyperodapedon," *Quart. Jour. Geol. Soc.* xxv. (1869), 138-152. *Sci. Mem.* iii.

"On a New Labyrinthodont (Pholiderpeton Scutigerum) from Bradford," *Quart. Jour. Geol. Soc.* xxv. (1869), 309-310. *Sci. Mem.* iii.

"On the Upper Jaw of Megalosaurus," *Quart. Jour. Geol. Soc.* xxv. (1869), 311-314. *Sci. Mem.* iii.

"Principles and Methods of Paleontology." (Written in 1865 as the Introduction to the Collection of Fossils at Jermyn Street.) *Smithsonian Report* (1869), 363-388. See p. 454 above (1865).

"On the Representatives of the Malleus and the Incus of Mammalia in the Other Vertebrata," *Proc. Zool. Soc.* (1869), 391-407. *Sci. Mem.* iii.

"Address to the Geological Society, 1869," *Quart. Jour. Geol. Soc.* xxv. (1869), 28-53. *Sci. Mem.* iii.

"On the Ethnology and Archæology of India." (Opening Address of the President, March 9, 1869.) *Jour. Ethn. Soc. of London,* i. (1869), 89-93. (Delivered March 9, 1869.) *Sci. Mem.* iii.

"On the Ethnology and Archæology of North America." (Address of the President, April 13, 1869.) *Jour. Ethn. Soc. of London,* i. (1869), 218-221. *Sci. Mem.* iii.

"On Hypsilophodon Foxii, a New Dinosaurian from the Wealden of the Isle of Wight" (1869), *Quart. Jour. Geol. Soc.* xxvi. (1870), 3-12. *Sci. Mem.* iii.

"Further Evidence of the Affinity between the Dinosaurian Reptiles and Birds" (1869), *Quart. Jour. Geol. Soc.* xxvi. (1870), 12-31. *Sci. Mem.* iii.

"On the Classification of the Dinosauria, with Observations on the Dinosauria of the Trias" (1869), *Quart. Jour. Geol. Soc.* xxvi. (1870), 32-50. *Sci. Mem.* iii.

"On the Ethnology of Britain," *Jour. Ethn. Soc. of London,* ii. (1870), 382-384. (Delivered May 10, 1870). *Sci. Mem.* iii.

" The Anniversary Address of the President," *Jour. Ethn. Soc. of London*, new series, ii. (1870), xvi-xxiv. (May 24, 1870). *Sci. Mem.* iii.

" On the Geographical Distribution of the Chief Modifications of Mankind," *Jour. Ethn. Soc. of London*, new series, ii. (1870), 404-412. (June 7, 1870.) *Sci. Mem.* iii.

" On a New Labyrinthodont from Bradford." With a Note on its Locality and Stratigraphical Position, by Louis C. Miall. *Phil. Mag.* xxxix. (1870), 385.

" Anniversary Address to the Geological Society, 1870," *Quart. Jour. Geol. Soc.* xxvi. (1870), 29-64. (*Paleontology and the Doctrine of Evolution*), *Collected Essays*, viii. 340. *Sci. Mem.* iii.

" Address to the British Association at Liverpool," *Brit. Assoc. Report*, xl. (1870), 73-89. *Collected Essays*, viii. *Sci. Mem.* iii.

" On the Milk Dentition of Palæotherium Magnum," *Geol. Mag.* vii. (1870), 153-155. *Sci. Mem.* iii.

" Triassic Dinosauria," *Nature*, i. (1870), 23-24. *Sci. Mem.* iii.

" On the Maxilla of Megalosaurus," *Phil. Mag.* xxxix. (1870), 385-386.

" On the Relations of Penicillium, Torula, and Bacterium," *Quart. Jour. Micr. Sci.* x. (1870), 355-362. (A Report by another hand of an Address given at the British Association, the views expressed in which were afterwards set aside.) *Sci. Mem.* iii.

" On a Collection of Fossil Vertebrata from the Jarrow Colliery, County of Kilkenny, Ireland," *Trans. Royal Irish Academy*, xxiv. (1871), 351-370.

" Yeast," *Contemporary Review*, December 1871. *Sci. Mem.* iii.

" Note on the Development of the Columella Auris in the Amphibia," *Brit. Assoc. Report*, 1874 (sect.), 141-142 ; *Nature*, xi. (1875), 68-69. *Sci. Mem.* iv.

" On the Structure of the Skull and of the Heart of

Menobranchus Lateralis," *Proc. Zool. Soc.* (1874), 186-204. *Sci. Mem.* iv.

" On the Hypothesis that Animals are Automata, and its History," *Nature*, x. (1874), 362-366. See also list of Essays.

" Preliminary Note upon the Brain and Skull of Amphioxus Lanceolatus " (1874), *Proc. Roy. Soc.* xxiii. (1875). *Sci. Mem.* iv.

" On the Bearing of the Distribution of the Portio Dura upon the Morphology of the Skull " (1874), *Prcc. Camb. Phil. Soc.* ii. (1876), 348-349. *Sci. Mem.* iv.

" On the Classification of the Animal Kingdom " (1874), *Jour. Linn. Soc. (Zool.)* xii. (1876), 199-226. *Sci. Mem.* iv.

" On the Recent Work of the *Challenger* Expedition, and its Bearing on Geological Problems," *Proc. Roy. Inst.* vii. (1875), 354-357. *Sci. Mem.* iv.

" On Stagonolepis Robertsoni, and on the Evolution of the Crocodilia," *Quart. Jour. Geol. Soc.* xxxi. (1875), 423-438. *Sci. Mem.* iv.

" Contributions to Morphology. Ichthyopsida. — No. 1. On Ceradotus Forsteri, with Observations on the Classification of Fishes," *Proc. Zool. Soc.* (1876), 24-59. *Sci. Mem.* iv.

" On the Position of the Anterior Nasal Apertures in Lepidosiren," *Proc. Zool. Soc.* (1876), 180-181. *Sci. Mem.* iv.

" On the Nature of the Cranio-Facial Apparatus of Petromyzon," *Jour. Anat. and Phys.* x. (1876), 412-429. *Sci. Mem.* iv.

" The Border Territory between the Animal and the Vegetable Kingdoms " (1876), *Proc. Roy. Inst.* viii. (1879), 28-34. *Macmillan's Magazine*, xxxiii. 373-384. *Sci. Mem.* iv.

" On the Evidence as to the Origin of Existing Vertebrate Animals," *Nature*, xiii. (1876), 388-389, 410-412, 429-430, 467-469, 514-516 ; xiv. (1876), 33-34. *Sci. Mem.* iv.

"The Crocodilian Remains in the Elgin Sandstones, with remarks on the Ichnites of Cummingstone," *Mem. Geol. Surv. of the United Kingdom,* Monogr. iii. 1877 (58 pp. and 16 plates). *Sci. Mem.* iv.

"On the Study of Biology," *Nature,* xv. (1877), 219-224 ; *American Naturalist,* xi. (1877), 210-221. *Sci. Mem.* iv.

"On the Geological History of Birds" (March 2, 1877), *Proc. Roy. Inst.* viii. 347. [The substance of this paper is contained in the *New York Lectures on Evolution,* 1876 ; see p. 440.]

"Address to the Anthropological Department of the British Association, Dublin, 1878. Informal Remarks on the Conclusions of Anthropology," *Brit. Assoc. Report,* 1878, 573-578. *Sci. Mem.* iv.

"On the Classification and the Distribution of the Crayfishes," *Proc. Zool. Soc.* (1878), 752-788. *Sci. Mem.* iv.

"On a New Arrangement for Dissecting Microscopes" (1878), the President's Address, *Jour. Quekett Micr. Club,* v. (1878-79), 144-145. *Sci. Mem.* iv.

"William Harvey" (1878), *Proc. Roy. Inst.* viii. (1879), 485-500. *Sci. Mem.* iv.

"On the Characters of the Pelvis in the Mammalia, and the Conclusions respecting the Origin of Mammals which may be based on them," *Proc. Roy. Soc.* **xxviii.** (1879), 295-405. *Sci. Mem.* iv.

"Sensation and the Unity of Structure of Sensiferous Organs" (1879), *Proc. Roy. Inst.* ix. (1882), 115-124. See also *Collected Essays,* vi. *Sci. Mem.* iv.

"The President's Address" (July 25, 1879), *Jour. Quekett Micr. Club,* v. (1878-79), 250-255. *Sci. Mem.* iv.

"On certain Errors respecting the Structure of the Heart, attributed to Aristotle" (1879), *Nature,* xxi. (1880), 1-5. See also *Science and Culture. Sci. Mem.* iv.

"On the Epipubis in the Dog and Fox," *Proc. Roy. Soc.* **xxx.** (1880), 162-163. *Sci. Mem.* iv.

"The Coming of Age of 'The Origin of Species'" (1880),

Proc. Roy. Inst. ix. (1882), 361-368.   See also *Collected Essays*, ii.   *Sci. Mem.* iv.

"On the Cranial and Dental Characters of the Canidæ," *Proc. Zool. Soc.* (1880), 238-288.   *Sci. Mem.* iv.

"On the Application of the Laws of Evolution to the Arrangement of the Vertebrata, and more particularly of the Mammalia," *Proc. Zool. Soc.* (1880), 649-662. *Sci. Mem.* iv.

"The Herring," *Nature*, xxiii. (1881), 607-613.   *Sci. Mem.* iv.

"Address to the International Medical Congress, London, 1881,—The Connection of the Biological Sciences with Medicine," *Nature*, xxiv. (1881), 342-346.   *Sci. Mem.* iv.

"The Rise and Progress of Paleontology," *Nature*, xxiv. (1881), 452-455.   *Sci. Mem.* iv.

"A Contribution to the Pathology of the Epidemic known as the 'Salmon Disease'" (February 21, 1882), *Proc. Roy. Soc.* xxxiii. (1882), 381-389.   *Sci. Mem.* iv.

"On the Respiratory Organs of Apteryx," *Proc. Zool. Soc.* (1882), 560-569.   *Sci. Mem.* iv.

"On Saprolegnia in Relation to the Salmon Disease," *Quart. Jour. Micr. Sci.* xxii. (1882), 311-333 (reprinted from the 21st Annual Report of H.M. Inspectors of Salmon Fisheries).   *Sci. Mem.* iv.

"On Animal Forms," being the Rede Lecture for 1883 ; *Nature*, xxviii. p. 187.

"Address delivered at the Opening of the Fisheries Exhibition at South Kensington, 1883."

"Contributions to Morphology. Ichthyopsida.—No. 2. On the Oviducts of Osmerus ; with Remarks on the Relations of the Teleostean with the Ganoid Fishes," *Proc. Zool. Soc.* (1883), 132-139.   *Sci. Mem.* iv.

"Oysters and the Oyster Question" (1883), *Proc. Roy. Inst.* x. (1884), 336-358.   *Sci. Mem.* iv.

"Preliminary Note on the Fossil Remains of a Chelonian Reptile, Ceratochelys Sthenurus, from Lord Howe's

Island, Australia," *Proc. Roy. Soc.* xlvi. (1887), 232-238. (Read March 31, 1887.) *Sci. Mem.* iv.

"The Gentians: Notes and Queries" (April 7, 1887), *Jour. Linn. Soc. (Botany)*, xxiv. (1888), 101-124. *Sci. Mem.* iv.

"Further Observations on Hyperodapedon," *Quart. Jour. Geol. Soc.* xliii. (1878), 675-693. *Sci. Mem.* iv.

"Owen's Position in the History of Anatomical Science," see p. 443.

APPENDIX IV[1]

HONOURS, DEGREES, SOCIETIES, ETC.

ORDER

Norwegian Order of the North Star, 1873.

DEGREES, ETC.

Oxford—Hon. D.C.L. 1885.
Cambridge—Hon. LL.D. 1879.
     Rede Lecturer, 1883.
London—First M.B. and Gold Medal, 1845.
    Examiner in Physiology and Comparative Anatomy, 1857.
    Member of Senate, 1883.
Edinburgh—Hon. LL.D. 1866.
Aberdeen—Lord Rector, 1872.
Dublin—Hon. LL.D. 1878.
Breslau—Hon. Ph.D. and M.A. 1861.
Würzburg—Hon. M.D. 1882.
Bologna—Hon. M.D. 1888.
Erlangen—Hon. M.D. 1893.

---

[1] This list has been compiled from such diplomas and letters as I found in my father's possession.

SOCIETIES—LONDON

Royal, 1851 ; Sec. 1872-81 ; Pres. 1883-85 ; Royal
  Society's Medal, 1852 ; Copley Medal, 1888 ; Darwin
  Medal, 1894.
Linnean, 1858 ; Linnean Medal, 1890.
Geological, 1856 ; Sec. 1859-62 ; Pres. 1869-70 ; Wollas-
  ton Medal, 1876.
Zoological, 1856.
Odontological, 1863.
Ethnological, 1863 ; Pres. 1868-70.
Anthropological Institute, 1870.
Medico-Chirurgical, Hon. Memb. 1868.
Medical, Hon. Memb. 1873.
Literary, 1883.
Silver Medal of the Apothecaries' Society for Botany, 1842.
Royal College of Surgeons, Member, 1862 ; Fellow, 1883 ;
  Hunterian Professor, 1863-69.
St. Thomas's Hospital, Lecturer in Comparative Anatomy,
  1854.
British Association for the Advancement of Science, Pres.
  1870 ; Pres. of Section D, 1866.
Royal Institution, Fullerian Lecturer, 1863-67.
British Museum, Trustee, 1888.
Quekett Microscopical Club, President, 1878-79.

SOCIETIES—PROVINCIAL, COLONIAL AND INDIAN

Dublin University Zoological and Botanical Association ;
  Corr. Member, 1859.
Liverpool Literary and Philosophic Society, Hon. Memb
  1870.
Manchester Literary and Philosophical Society, Hon.
  Memb. 1872.
Odontological Society of Great Britain, 1862.
Royal Irish Academy, Hon. Memb. 1874.

Historical Society of Lancashire and Cheshire, Hon. Memb.
     1875.
Royal Society of Edinburgh, British Hon. Fellow, 1876.
Glasgow Philosophical Society, Hon. Memb. 1876.
Literary and Antiquarian Society of Perth, Hon. Memb.
     1876.
Cambridge Philosophical Society, Hon. Memb. 1871.
Hertfordshire Natural History Society, Hon. Memb.1883.
Royal College of Surgeons of Ireland, Hon. Memb. 1886.
New Zealand Institute, Hon. Memb. 1872.
Royal Society of New South Wales, Hon. Memb. 1879,
     Clarke Medal, 1880.

FOREIGN SOCIETIES

International Congress of Anthropology and Prehistoric
     Archæology, Corr. Memb. 1867.
International Geological Congress (Pres.) 1888.

America

Academy of the Natural Sciences of Philadelphia, Corr.
     Memb. 1859 ; Hayden Medal, 1888.
Odontographic Society of Pennsylvania, Hon. Memb. 1865.
American Philosophical Society of Philadelphia, 1869.
Buffalo Society of Natural Sciences, Hon. Memb. 1873.
New York Academy of Sciences, Hon. Memb. 1876.
Boston Society of Natural History, Hon. Memb. 1877.
National Academy of Sciences of the U.S.A., Foreign
     Associate, 1883.
American Academy of Arts and Sciences, Foreign Hon.
     Memb. 1883.

Austria-Hungary

Königliche Kaiserliche Geologische Reichsanstalt (Vienna),
     Corr. Memb. 1860.
K.K. Zoologische-botanische Gesellschaft in Wien, 1865.

Belgium

Académie Royale de Médecine de Belgique, 1874.
Société Géologique de Belgique, Hon. Memb. 1877.
Société d'Anthropologie de Bruxelles, Hon. Memb. 1884.

Brazil

Gabineta Portuguez de Leitura em Pernambuco, Corr.
Memb. 1879.

Denmark

Royal Society of Copenhagen, Fellow, 1876.

Egypt

Institut Egyptien (Alexandria), Hon. Memb. 1861.

France

Société Impériale des Sciences Naturelles de Cherbourg,
Corr. Memb. 1867.
Institut de France; "Correspondant" in the section of
Physiology (succeeding von Baer), 1879.

Germany

Microscopical Society of Giessen, Hon. Memb. 1857.
Imperialis Academia Caesariana Naturae Curiosorum
(Dresden), 1857.
Imperial Literary and Scientific Academy of Germany,
1858.
Royal Society of Sciences in Göttingen, Corr. Memb. 1862.
Royal Bavarian Academy of Literature and Science
(Munich), For. Memb. 1863.
Royal Prussian Academy of Sciences (Berlin), 1865.
Medicinisch-naturwissenschaftliche Gesellschaft zu Jena,
For. Hon. Memb. 1868.
Geographical Society of Berlin, For. Memb. 1869.
Deutscher Fischerei-Verein, Corr. Memb. 1870.

Berliner Gesellschaft für Anthropologie, Ethnologie, und
Urgeschichte, Corr. Memb. 1871.
Naturforschende Gesellschaft zu Halle, 1879.
Senkenbergische Naturforschende Gesellschaft (Frankfurt
a/M.), Corr. Memb. 1892.

Holland

Dutch Society of Sciences (Haarlem), For. Memb. 1877.
Koninklyke Natuurkundige Vereenigung in Neder-
landisch-Indie (Batavia), Corr. Memb. 1880.
Royal Academy of Sciences (Amsterdam), For. Memb.
1892.

Italy

Società Italiana di Antropologia e di Etnologia, Hon.
Memb. 1872.
Academia de' Lincei di Roma, For. Memb. (supplementary),
1878, ordinary, 1883.
Reale Academia Valdarnense del Poggio (Florence), Corr.
Memb. 1883.
Società dei Naturalisti in Modena, Hon. Memb. 1886.
Società Italiana delle Scienze (Naples), For. Memb. 1892.
Academia Scientiarum Instituti Bononiensis (Bologna),
Corr. Memb. 1893.

Portugal

Academia Real das Sciencias de Lisboa, For. Corr. Memb.
1874.

Russia

Imperial Academy of Sciences (St. Petersburg), Corr.
Memb. 1865.
Societas Caesarea Naturae Curiosorum (Moscow), Ordinary
Member, 1870, Hon. Memb. 1887.

Sweden

Societas Medicorum Svecana, Ordinary Memb. 1866.

ROYAL COMMISSIONS

T. H. Huxley served on the following Royal or other Commissions :—

1. Royal Commission on the Operation of Acts relating to Trawling for Herrings on the Coast of Scotland, 1862.
2. Royal Commission to inquire into the Sea Fisheries of the United Kingdom, 1864-65.
3. Commission on the Royal College of Science for Ireland, 1866.
4. Commission on Science and Art Instruction in Ireland, 1868.
5. Royal Commission upon the Administration and Operation of the Contagious Diseases Acts, 1870-71.
6. Royal Commission on Scientific Instruction and the Advancement of Science, 1870-75.
7. Royal Commission on the Practice of subjecting Live Animals to Experiments for Scientific Purposes, 1876.
8. Royal Commission to inquire into the Universities of Scotland, 1876-78.
9. Royal Commission on the Medical Acts, 1881-82.
10. Royal Commission on Trawl, Net, and Beam Trawl Fishing, 1884.

INDEX

THE END

Printed by R. & R. CLARK, LIMITED, *Edinburgh.*

9 781108 040488